W9-BKB-343

THE MYTH OF PROGRESS

THE MYTH OF PROGRESS

Toward a Sustainable Future

TOM WESSELS

UNIVERSITY OF VERMONT PRESS
Burlington, Vermont

Published by University Press of New England
Hanover and London

University of Vermont Press
Published by University Press of New England,
One Court Street, Lebanon, NH 03766
www.upne.com

Printed in the United States of America
10 9 8 7

Library of Congress Cataloging-in-Publication Data

Wessels, Tom, 1951–
The myth of progress : toward a sustainable future / Tom Wessels.
p. cm.
Includes index.
ISBN-13: 978–1–58465–495–7 (cloth : alk. paper)
ISBN-10: 1–58465–495–3 (cloth : alk. paper)
1. Ecology. 2. Environmental policy. 3. Sustainable development. I. Title.
QH541.W42 2006
577—dc22 2006016895

University Press of New England is a member of the Green Press Initiative. The paper used in this book meets their minimum requirement for recycled paper.

Acknowledgments

The idea for this book was sparked by a comment one of my students made in my principles of ecology course at Antioch. Although *The Myth of Progress* evolved into something quite different from its original conception, thanks to Susie Caldwell Rinehart for planting the seed. Thanks also to Jared Volpe for doing the initial research on indicators of progress and to my daughter, Kelsey Wessels, who served as my able research assistant throughout the development of the manuscript. Special thanks to my wife, Marcia Wessels, for her careful reading, editing, and constant support during the writing of the manuscript. I am grateful to my colleague Steve Chase for his extremely helpful suggestions on the book's content and organization. Steve's guidance was essential in more effectively merging my scientific background with current socioeconomic trends. Finally, I am indebted to Helen Whybrow, John Elder, and Tom Hudspeth for their very positive endorsements of this project and the staff at University Press of New England who have eagerly supported this project. Without their collective efforts this book might not have been published.

To the memory of
Donella Meadows
and her inspirational work

Contents

Prologue

I grew up during the 1950s in a suburban development spawned by American optimism following World War II. Like so many of these developments, similar looking houses were evenly spaced on tidy one-half acre lots. Luckily for me, across the street from our house was an intact forest probably 70 acres in size. From my current ecological perspective, 70 acres doesn't seem like much, but back then it was an extensive wilderness. At the age of five, I started to walk to school through those woods, and for the next seven years the majority of my free time was spent in that forest building forts, climbing trees and cliffs, exploring boulder caves, and watching fat pollywogs grow into frogs. My early roots as an ecologist and natural historian were put down in those woods. My first experience of "development" occurred there too.

It was during the spring of my sixth-grade year, just as the leaves were opening on the maples, that two of my close friends and I were walking home from an after-school baseball game late in the afternoon. About five minutes into the woods we stumbled upon a brand new bulldozed road that had been punched right into the center of the forest. Shocked by this intrusion into our sanctuary, we followed the road a short distance and to our dismay found that the bulldozer had run right over one of our forts. Without saying a word we became of one mind and started to run down this new gash farther into the forest. The road ended in a clearing where the bulldozer had toppled a number of hoary-barked, old red maples and pushed them to one side. In the middle of the clearing sat the unattended bulldozer—its operator apparently having finished work for the day.

We were outraged; without discussing it we started hurling stones at the dozer's windows. Once we had cracked all the windows, we got some sharp sticks and punched holes throughout the leather seat cover and pulled out chunks of foam. Then into every gear that we could reach we jammed rocks, and with stout branches we pried off some hydraulic hoses. Finally we poured dirt, exposed by the dozer's tracks, into its gas tank. We never questioned what we were doing, and it is the only time in my life that I vandalized someone else's property. But to the three of us, those woods were our home. The bulldozer had invaded the sanctity of that home, and we fought back.

Just about every person I have met who is close to my age has a similar experience of the loss of a special childhood place. Prior to the nineteenth century, the vast majority of human beings lived in landscapes where their ancestors had existed for generation after generation. In this way, people were intimately tied to their place. It's a very recent phenomenon that landscapes to which people were once connected have become smothered by development—growth that we are told is a sign of progress. But is progress truly possible if its wake continually generates loss—loss of connections to place and community, loss of clean air and water, loss of other species who are truly part of our ancestral family tree?

Although we probably did some serious damage to the dozer, within a few short months the entire forest was gone and soon replaced by a hundred new homes. Since that time I have learned that violence doesn't accomplish anything. So my efforts today are focused on educating people about the wonders of the natural world, hoping that this will foster stronger connections, stewardship, and care of our biological heritage—a heritage that has taken more than three billion years to develop and on which our existence is completely dependent.

Now I am taking my experience as educator, ecologist, and natural historian and applying it to an examination of our notions of progress. I hope that just as I have helped people to see and experience natural history in a new way, I can offer an alternative view of our current socioeconomic system. I am not an experienced economist, political scientist, or sociologist. But as an ecologist I am well versed in the foundational laws that govern all complex systems, and a socioeconomic system is a complex one. It is within this context that I believe I can offer a sobering view of our current march toward progress.

Because my critique of our reigning notions of progress is scientifically based, scientific terms are used in this book. Those terms are listed in the glossary and appear in italics when first introduced in the text.

Introduction

"Economic growth is key to environmental progress." —George W. Bush, 2/14/02

The above statement is an excerpt from a speech that President Bush gave on Valentine's Day. I heard it while listening to National Public Radio's *All Things Considered* news program as I was driving home from Antioch New England Graduate School, where I teach. I found the President's statement so provocative that I had to pull over so I could write it down. As an ecologist, it's evident to me that economic growth, as well as its associated ever-increasing extraction of resources and waste generation, is primarily responsible for the environmental problems that we are witnessing today. How can it be that a world leader would suggest that the solution to our environmental problems is more economic growth? It's possible that President Bush was being disingenuous in his statement. But it is equally possible that his assessment is the result of his paradigmatic view of progress. If we remove the word "environmental," the President's statement becomes "Economic growth is key to progress"—an opinion that is shared by the majority of people and policy makers today.

Paradigm is a word derived from the Greek *paradeigma*, which means to show side by side. This sounds like a rather innocuous thing to do, but with our more modern concept of paradigm such comparisons can be powerful and at times volatile. In our current use, a paradigm represents a core belief that dramatically structures our worldview. It is a lens through which all of our perceptions and thoughts are strongly filtered. Wherever differing paradigms intersect, there will at least be debate, often confrontation, and sometimes violence. The contentiousness that so infuses the issue of abortion is the result of differing paradigmatic views of what constitutes life; the clash of these differing views often sparks violent acts.

In very powerful ways, we are shrouded and entrapped within the paradigms that we accept—and this acceptance is often an unconscious act. Reigning cultural paradigms can be passed from generation to generation, and if they aren't challenged, they are simply accepted as truth. To change one's paradigm is perhaps the most difficult of challenges, because it often requires turning one's world inside out. People have killed other people over challenged paradigms. And people have even sacrificed their children and families rather than sacrifice their core beliefs. To change one's paradigm is a dramatic event.

This book is a critique of our reigning paradigm of progress—that in order to progress we need to keep growing the economy. I need to distinguish here between economic growth and economic development. Economic growth is predicated on increasing consumption of resources, whereas economic development can occur without increased consumption. Economic development can be encouraged through value-added activities. In Vermont, dairy farming is the foundation of the agricultural economy. But due to the nature of the state's mountainous landscape and narrow valleys, dairy farms are restricted in size. This restriction makes it difficult for Vermont farms to compete with large western dairies that can have thousands of cows. So rather than just selling raw milk, a number of Vermont farms have added value to their operations by turning the milk into cheese or yogurt. They use the same amount of resources, but they add value by further developing their milk products. My critique is focused solely on economic growth.

Previous books have criticized this notion of progress. In 1971 economist Nicholas Georgescu-Roegen, in his book *The Entropy Law and the Economic Process*, attacked the idea of unlimited growth as being scientifically unfounded because it ignored the second law of thermodynamics. The following year, in *The Limits to Growth*, Donella Meadows and associate systems scientists from the Club of Rome showed through computer modeling that unlimited growth was not sustainable. In 1977 Herman Daly came forth with *Steady-State Economics*, which developed an alternative model to economic progress. More recently Daly partnered with John Cobb in writing *For the Common Good*, which furthers his previous work in sustainable economies. Then, in the 1990s, works by Hazel Henderson, Paul Hawken, and David Korten developed the model of sustainable economies that functioned under principles found in biological systems. *The Myth of Progress* builds on this developing conversation, but it is distinguished from these previous works by its scientific examination of a number of interrelated, foundational laws (what I call the laws of sustainability) that govern the behavior of all *complex systems*.

Complex systems science—formerly coined as *chaos theory*—is a scientific perspective of the world that arose during the second half of the twentieth century. Today it is challenging Western science's long-held, linear, reductionistic perspective, which has framed scientific and cultural thought since the time of Galileo. Since progress is a direct outcome of our socioeconomic system—a system that is complex and decidedly not linear—understanding the principles that govern how complex systems function is necessary to evaluating our current beliefs about progress.

In the 1965 edition of *Webster's New Collegiate Dictionary*, progress is defined as the gradual betterment of mankind. Today, I would guess that "humankind" has replaced "mankind." But in either case, are we progressing? And what does the future hold for the progress of humanity? The accepted answer to both of these questions in our post-industrial society is an unqualified "yes"—we are currently progressing, and the future will see us progressing ever further.

Our reigning paradigm of progress—one that drives global policy decisions—was birthed in the nineteenth century. It is embraced, and usually unquestioned, by the majority of people in our culture. It serves to guide policy decisions—from town planning boards, to state and federal agencies, to the Oval Office, and is the foundation of the push for globalization that we are witnessing today. But what if this paradigm of progress, one that was forged long before our understanding of how complex systems function, is flawed? What if it is an ill-founded belief—a myth?

Myth, as my friend Richard Thompson pointed out while we were hiking in the White Mountains, actually has two definitions. One defines myth as a legendary story that at its core unveils some truth. These myths are the type that Joseph Campbell so eloquently presented. Another definition of myth holds that it is an accepted belief that is fallacious.

It is within this second context that "myth" is used in this book.

The ruling paradigm rests on the idea that for more than a century, things have dramatically improved for people on this planet—and will continue to improve—as economic systems keep expanding. So as President Bush stated, continued economic growth equates to progress. But a number of economists including Georgescu-Roegen, Daly, and Kenneth Boulding (once president of the American Economic Association), plus such scientists as Donella Meadows and ecologist Paul Ehrlich have questioned this perspective. They suggest that at some point economic expansion will outstrip both the availability of resources and the ability of the Earth to absorb the waste products generated by this growth. As this point approaches progress will slow, eventually cease, and then retreat. Neoclassical economists always counter this concern with the following: As we continue to develop new technologies and harness ever-increasing amounts of energy, we will be able to stay well ahead of resource depletion and be able to handle waste generation, and as a result, progress will continue. As we will see, scientific laws suggest otherwise.

Possibly the most ardent supporter of continuous growth was economist Julian Simon, who asserted, "We now have in our hands—in our libraries, really—the technology to feed, clothe, and supply energy to an ever-growing population for the next 7 billion years."[1] When Simon made this assertion, the global population was increasing at around 1.6 percent a year. Al Bartlett, a retired University of Colorado physicist, decided to calculate a sustained rate of annual growth of the human population at just 1 percent. At this rate it would take only seventeen thousand years for the number of people to equal the number of all the atoms in the universe.[2] Obviously Simon's assertion regarding global population growth is not only impossible, but also extremely naive. Yet he is revered in the neoclassical economic community.

The fervent faith that neoclassical economists place in unlimited growth is based on and supported by two things. The first are indicators of what I call "material progress." These indicators track such things as increasing GDP (gross domestic product), increases in per capita income, and longer life expectancies. The second is a linear model of progress that doesn't take into account iterative feedback that can dramatically change how a complex system behaves. This linear view fosters predictability—that the system will function in the future just as it has in the past. *The Myth of Progress* is a critique of the failings of this linear view, but first let's look more closely at two indicators used to support the current notion of progress.

Increasing gross domestic product is the most broadly accepted indicator of progress. GDP is an annual summation of all the monetary transactions that occur within an economy. If more money changes hands, the economy is more robust, and therefore things must be better for people within that economy. But the GDP has a number of major flaws as an indicator of progress. A critical one is that money spent to tackle social, medical, or environmental problems is registered as a boost for GDP. Most people would agree that increasing levels of air pollution would not indicate progress. But as air pollution goes up, so do medical costs associated with asthma, emphysema, and heart disease. The money spent to deal with these increased medical problems is registered as growth of the GDP, as is the money spent to clean up environmental messes like superfund sites, or money spent to fight crime, drug addiction, and terrorism. Costs associated with declines in social or environmental quality actually boost GDP.

A second flaw is that activities that don't involve the exchange of money are not part of the GDP. How important is the work that takes place within families or volunteerism within communities? I think most people would say that intact families and vibrant communities would be strong indicators of progress. Yet these never figure into GDP, and even worse, when crime or drug use isolate people in communities, or when parents need to work longer hours (resulting in less time for family interaction), the GDP goes up, not down.

Finally, rising GDP doesn't lift all boats. In the 1980s as the U.S. GDP rose 20 percent, the richest 1 percent of Americans increased their wealth by 60 percent whereas the majority of Americans saw little gain, and the bottom 40 percent of Americans got poorer.[3] Since the 1980s middle-class families have had to work longer hours simply to maintain their 1980 standard of living as the GDP experienced its steepest growth.[4] These trends continue today. In 2003 the top 10 percent of Americans saw their salaries increase 1.1 percent, while the middle class saw their earnings drop 0.1 percent, and the bottom 10 percent lost 1.2 percent of their salaries.[5] In 1979 the mean salary for individuals in the top 10 percent was 3.7 times greater than those for individuals in the bottom 10 percent. In 2003 that difference had grown to 4.7 times greater.[6]

In response to the GDP's problems as a flawed indicator of progress, Mark Anielski and Jonathan Rowe of *Redefining Progress* developed a new indicator, the GPI or genuine progress indicator. The GPI adjusts for the social, environmental, family, and medical costs mentioned above. Tracking both the

GDP and GPI since 1950 shows that they grew in tandem until about 1980. Then the GDP grew more steeply while the GPI started to fall, and continues to fall as social, health, and environmental problems grow more pervasive.[7] This might explain why a March 1994 *Business Week*/Harris poll found that 70 percent of the public was gloomy about the future when the U.S. economy and GDP were booming and per capita income (the average income of all wage earners) was increasing.[8] As the level of affluence of the wealthy increases, it can pull up per capita income figures even as the majority of wage earners' incomes stagnate or fall. The general experience for the average citizen in 1994 was that things didn't seem to be getting better. I would imagine that this perspective has only strengthened during our current "war on terror"—a war that has no apparent end in sight.

Better physical health, often measured in increasing life span, is also commonly used as an indicator of progress. But what are we to make of the following facts: heart disease is the number one cause of premature death in this country; suicide ranks eleventh,[9] with an estimated three students in the average high-school classroom having attempted to take their own lives[10]; and depression is the fourth leading cause of disability in the developed world.[11] Asthma, obesity, and type 2 diabetes—all diseases that have a strong environmental component—have dramatically increased over the past two decades. Between 1980 and 1996 the number of asthma attacks occurring in a twelve-month period increased 73.9 percent.[12] In 1985 only eight states reported obesity rates in the high range of 10 to 14 percent of their population. By 2000, twenty-two states reported obesity rates of greater than 20 percent of their population—more than a four fold increase of obesity in just fifteen years.[13] With the increasing rate of obesity, type 2 diabetes has also increased and now affects 6.5 percent of Americans.[14] Medical technology—a form of material progress—is allowing people to live longer, but these statistics do not suggest that people are living healthier, better lives.

Unipolar depression—a modern disease that has not been confirmed in nonindustrial societies—is at epidemic levels in industrialized nations.[15] During the last fifty years, the average age of onset of unipolar depression has dropped from the late forties to the late twenties.[16] More than 330 million people worldwide suffer from this disease, and by the year 2020 it is expected to be the second leading cause of disability in the world.[17] It is interesting to note that the Amish, who embrace a simple, community-based lifestyle, have rates of unipolar depression five times lower than their surrounding more affluent neighbors.[18] Simply looking at longevity or affluence as indicators of progress is much too simplistic. It wouldn't be wrong to speculate that something is amiss when the leading causes of death and disability in the modern world are diseases—unheard of in nonindustrial cultures—that attack the heart and the mind.

Although I have presented a number of figures in the last four paragraphs, this book will not be a compilation of endless facts and statistics. There are two reasons for this. The first is that facts and statistics do not necessarily represent truth. They can be intentionally manipulated to support any argument. This is why scientific research is supposed to be peer-reviewed. The generation of facts is directly linked to researchers' perspectives, the kinds of questions they ask, how they gather their data, and the statistical analyses they choose to use. As such, facts should be suspect until it can be seen how they were derived. Let me give an example.

In his book, *The Skeptical Environmentalist*, Bjørn Lomborg writes:

> Discussing forests, Worldwatch Institute categorically states that the world's forest estate has declined significantly in both area and quality in recent decades. As we shall see in the section on forests, the longest data series from the UN's FAO show that global forest cover has *increased* from 30.04 percent of the global land area in 1950 to 30.89 percent in 1994, an increase of 0.85 percentage points over the last 44 years.[19]

As a forest ecologist, I was intrigued by Lomborg's assertion that forest coverage expanded during a period of time that witnessed increasing rates of clearcutting and forest fragmentation in the tropics and other parts of the world. The facts he states come from the United Nation's Food and Agricultural Organization and are based on the following definition of forest: A forest is any site that is greater than 0.5 hectares in size, with trees that reach a minimum of five meters in height and a crown coverage that exceeds 10 percent of the area. This means that any site slightly larger than an acre, with widely spaced sixteen-foot tall trees that at midday shade only 11 percent of the ground, is a forest. Under this definition many suburban lawns, golf courses, and city parks would be considered forests. But that's not all. Reading the FAO's guidelines further I found that tree plantations, tree nurseries, and even *recent clearcuts* that don't meet the above standards also fall under the FAO's definition of a forest if in time they will grow sixteen-foot tall trees.[20] How does this definition of forest match yours?

Lomborg never defines what he means by "forest." So if readers don't take the time to examine the endnotes, and in this case go to the FAO's web site to check how the above facts were derived, they could easily think that healthy, intact, forest cover is increasing globally. If Lomborg had focused on global trends regarding healthy, intact forests, his facts would have supported the Worldwatch Institute's assertion forest area and quality have actually declined.

Another reason that this book doesn't focus on statistics is that it is important to understand the underlying principles governing complex systems and not end up lost in a sea of facts. *The Myth of Progress* guides readers to an understanding of a few critical, underlying, scientific laws that govern all complex systems—laws that our reigning paradigm of progress completely ignores. These laws include the law of limits to growth, the second law of thermodynamics, and the law of self-organization in complex systems. Yet before we examine these three laws of sustainability, it is necessary to understand the differences between linear and complex systems, and how these differences relate to the myth of progress.

Notes

1. Norman Myers and Julian Simon, 1994. *Scarcity or Abundance: A Debate on the Environment* (New York: W. W. Norton), 65.

2. A. Bartlett, 1997. "The Exponential Function, XI: The New Flat Earth Society." *Focus* (Carrying Capacity Network) vol. 7, no. 1:34–36.

3. C. Cobb et al., "The GDP Is Up, Why is America Down." *Atlantic Monthly*. October 1995: 72. See <http://www2.theAtlantic.com/atlantic/xchg/circ/back.htm>.

4. Ibid., 72.

5. "Wages Increase for Wealthy." *Brattleboro Reformer*, 12 February 2004, p. 15.

6. Ibid., 15.

7. M. Anielski and J. Rowe, 1999. *The Genuine Progress Indicator—1998 Update* (San Francisco: Redefining Progress), p. 4 <http://www.rprogress.org/newpubs/1999/gpi1999_execsum.pdf>.

8. See C. Cobb et al.

9. National Institute of Mental Health. Suicide Facts. See <http://www.nimh.nih.gov/suicideprevention/suifact.cfm>.

10. American Association of Suicidology Youth Suicide Fact Sheet. See <http://www.suicidologyt.org>.

11. R. Glass, 1999. "Treating Depression as a Recurrent or Chronic Disease." *Journal of the American Medical Association* vol. 281, no. 1:83.

12. D. Mannino et al., 2002. Surveillance for Asthma—United States, 1980–1996. Center for Disease Control. See <http://www.cdc.gov/mmwr/preview/mmwrhtml/ss5101a1.htm>.

13. A. Mokdad, et al., 2001. "The Continuing Epidemics of Obesity and Diabetes in the United States." *Journal of the American Medical Association* vol. 286, no. 10:1197.

14. National Diabetes Information Clearinghouse. See <http://diabetes.niddik.nih.gov/dm/pubs/statistics/index.htm#9>.

15. R. Wright, 1995. "The Evolution of Despair." *Time* vol. 146, no. 9:51. See <wysiwyg://83/http://www.time.com/time/m...e/archive/1995/950828/950828.cover.html>.

16. Myrna Weissman et al., 1996. "Cross-national Epidemiology of Major Depression and Bipolar Disorder." *Journal of the American Medical Association* vol. 276: p. 295.

17. 1998. "Spirit of the Age." *The Economist* vol. 349, no. 8099: 113.

18. See R. Wright, p. 52.

19. B. Lomborg, 2001. *The Skeptical Environmentalist: Measuring the Real State of the World* (Cambridge: Cambridge University Press), 13.

20. Food and Agriculture Organization of the United Nations. 2000. The Forest Resources Assessment Programme. See <http://www.fao.org/documents/show_cdr.asp?vrl_file=/docrep/007/ae217e/ae217e00.htm>.

THE MYTH OF PROGRESS

1

THE MYTH OF CONTROL

Complex versus Linear Systems

"The *control of nature* is a phrase conceived in arrogance, born of the Neanderthal age of biology and philosophy . . ." —Rachel Carson[1]

Chaos

My first exposure to computers based on microchip technology was in 1980. Jimmy Carter had just lost the presidency to Ronald Reagan. An evening news report showed Carter at home writing his memoirs on a "word processor," which if I remember correctly looked like an early Apple computer. Just eight years earlier I had been running punch cards through a vacuum tube–run IBM 360 mainframe computer that took the space of a small room at the University of New Hampshire's computer center. At that time the future of computers appeared to be in large mainframes like this one, which serviced a host of programmers; computers would be the domain of the highly trained. Who could have imagined that in less than a decade new technology would change the field so dramatically? As I watched Carter typing away on his personal computer, I realized that in eight short years all of my computer training had become obsolete. Yet the now-antiquated mainframe computer did generate some startling discoveries, possibly the most important being chaos theory.

It was on a Royal McBee, a vacuum tube–run computer in the early 1960s, that Edward Lorenz, a research meteorologist at MIT, inadvertently stumbled upon a finding that shook the very paradigmatic foundations of Western science. Lorenz's Royal McBee would look like a prehistoric dinosaur next to today's computers—a huge mass of tubes and wires that rattled loudly while operating. Although more than a hundred times as large as a personal computer, it had thousands of times less "brain power." Yet it could do something that people couldn't—it could execute millions of calculations in a relatively short span of hours.

Lorenz had been attracted to weather as a child and followed this interest to one of the premier research institutions in the world. Unlike astronomy—a physical science that could make fairly accurate long-term predictions regarding eclipses or the return of comets—meteorology had progressed little through the twentieth century in terms of accurately projecting the weather just a few days hence. Lorenz hoped to change that. Within his Royal McBee he created a virtual weather system. Through the coupling of twelve equations that related such things as pressure to wind direction or temperature to pressure, he produced a computerized system that mimicked the weather.[2] He hoped that he would be able to glean repeated patterns from his virtual system that could be applied to improving real weather forecasting.

During the winter of 1961 he stopped the computer in the midst of one of its runs to double check a weather sequence in his virtual world. He typed in the numbers from his printout at the point where he wanted to restart the run and left his office to let the McBee rattle away. Upon his return he was shocked to find that the new run, after just a few cycles, had totally diverged from the original run. Because each run started at the same point and followed the same laws as prescribed by his programmed equations, both runs should have been identical.

Lorenz double-checked the number he inputted to start the second run with those on the first printout. It was this comparison that led Lorenz to a starling discovery. For the second run he had entered the number 0.506, a rounded down version of the printout's number—0.506127. The two numbers differed by only 0.000127—a little more than one ten-thousandth.[3] Based on the well-established scientific notion—*proximate knowledge of initial conditions*—such a small change shouldn't have affected the outcome of the run. It was known in the scientific community that absolutely accurate measurements of anything were not possible. But having a close measurement of initial conditions—proximate knowledge—was fine for making future predictions due to *convergence*, a situation where minor perturbations in a system tend to cancel each other out, allowing the system to function in predictable ways. Under this paradigm, if you are a little off at the start in your measurements, it only means that you will be a little off at the end.

This concept is borne out in a scene from the movie *Apollo 13*. As the disabled space capsule carrying Jim Lovell and crew approaches the Earth, they have to readjust their angle of descent. If the angle too steep they will burn up; if it's too shallow, they will bounce off the Earth's atmosphere, never to return. They need to readjust their descent pattern by using thrusters for a prescribed period of time while holding the capsule in a fixed position. Of course the timing of the thruster use and the holding steady of the capsule couldn't be accomplished with absolute accuracy. It didn't matter, though, because if their initial conditions of thrust and position were proximate, it would be enough for a successful descent—one they accomplished.

What Lorenz saw in the second run of his Royal McBee made him realize that not only would long-term accurate weather prediction be unattainable, but more importantly the notion of proximate knowledge of initial conditions was flawed. Slight alterations at the start of a system could indeed dramatically alter its future behavior. This would come to be known as the *butterfly effect*, after a hypothetical scenario. The beat of a butterfly's wings in Asia creates minor air movements, initiating a long string of events that cascade through the meteorological system, eventually generating a powerful storm in the United States. As such, systems like Lorenz's virtual weather weren't predictable. Lorenz shared his findings with colleagues who, being steeped in more than two centuries of Newtonian physics and believing in the rightful place of proximate knowledge of initial conditions, rejected the finding. Later they would come around to call these systems chaotic. The naming of such systems as chaotic and the study of them as chaos theory show just how embedded in a linear paradigm western science was—a paradigm imbedded in predictability, spawned by the work of Galileo and Descartes, and later codified by Newton's grand synthesis.

Linear Systems

For almost 2,000 years prior to the seventeenth century, Aristotle's ideas formed the very foundation for Western natural science. At the heart of Aristotle's work was change. He saw matter and form as inextricably linked by a dynamic, developmental process of change that he labeled *entelechy*—self-completion. The matter of a rotting log is transformed into a fungus, or milkweed leaves consumed by a caterpillar become transformed into a monarch butterfly. For Aristotle, to develop an understanding of the world, it was necessary to comprehend how change was continuously restructuring matter into new forms. Through this approach to studying the natural world, process and pattern became far more important than the material of which something was composed. As we will see, Aristotle's underlying notions are strikingly similar to modern complex systems science.

In the early seventeenth century a new scientific paradigm emerged from the work of Galileo and later René Descartes. Both men advanced the idea that the world was comprised of matter in motion. The best approach to understanding the nature of matter in motion was to reduce problems to simple terms that could be analyzed and solved through simple mathematical equations. This approach gave rise to *reductionism*—to understand how something worked, it had to be taken apart, and its parts studied at increasingly smaller levels. Using this approach Galileo was able to predict movements of planets in somewhat accurate ways.

Descartes took reductionism even further. He was intrigued by machines that were prevalent in Europe in the early seventeenth century, such as wind-up clocks and various wind and water mills. These devices had a profound impact on his thinking, causing him to view the natural world as being composed solely of machines. "We see clocks, artificial fountains, mills and other similar machines which, though merely man-made, have nonetheless the power to move by themselves in several different ways . . . I do not recognize any difference between the machines made by craftsmen and the various bodies that nature alone composes."[4]

He knew that a machine could be understood if one knew all the parts and the sequence in which they interacted. In order to know the parts, one needed to take apart the machine—reductionism. Since each part would drive another that would in turn drive yet another part, the machine represented a *linear system*. A linear system does not mean that the system can't run in a cyclic fashion like the hands of a clock. But it does mean that the system can't feed back on itself. In a linear system each part works in a lockstep way with the other parts, so that the system always follows the exact same sequence of interactions between the parts. In this way a linear system is extremely predictable, and as such, controllable. After so many ticks of the gears, the minute and hour hands of a clock will have moved only so far. For Descartes the sum of a machine's parts simply equaled the whole. What is lost in this paradigmatic view of the world is that the whole may be much more than the sum of its parts. More than any other scientist of his time, René Descartes changed the paradigmatic nature of Western science—from Aristotle's holistic view to a reductionistic, linear view that focused on the parts rather than the whole. But it was Sir Isaac Newton who codified linear science with his study of mechanics.

I remember my high-school physics class and all sorts of experiments that we conducted to confirm Newton's laws of mechanics, such as F = MA—force equals mass times acceleration. However, to get close to proving these laws, we had to use air machines to reduce friction as much as possible, or drop high-density items rather than low-density ones. The dropping of baseballs from our third story high school windows worked well to confirm Newton's laws. But if we had dropped dried leaves—impacted by wind currents and their own tumbling behavior—the equations wouldn't have worked at all. The failure of Newtonian mechanics in certain circumstances was never discussed in this class. Instead Newton's work was always shown to display the hallmark of science—predictability.

Descartes's and Newton's approaches to scientific inquiry have proven very powerful in various branches of science over the last four centuries, but their approaches have huge deficiencies when applied to the complex natural systems that surround us—biological, geological, meteorological—as well as human-generated systems, such as an economy. These systems all have parts that can interact with other parts in different ways at different times, allowing these systems to loop, or feed back on themselves. It is this process of feedback that creates a number of profound differences between complex and linear systems.

Complex Systems

Because of a complex system's ability to feed back on itself, it loses quickly the inherent predictability of a linear system. Each morning millions of people in this country commute to work, often driving cars into urban areas. For these commuters it is not possible to predict exactly how long the commute will take. The choked highways generate their own feedback. A driver leaves home in good spirits, but finds he is becoming anxious at the somewhat slower pace of the commute. He becomes more aggressive than usual and cuts into a lane too close to the car behind him. That individual brakes unexpectedly and gets rear-ended by the car behind him. One lane of traffic is stopped and backs up the highway for miles.

In this case the slower commute fed back on itself, creating conditions that further slowed the drive. This is called *positive feedback* because it forces the system to keep moving in the direction it started—slow to slower.

The same is true for weather. Last night our weather report predicted two to four inches of snow by morning. It's now 9 A.M., partially sunny, and no snow has fallen. This morning's weather report talks only of flurries this afternoon. How could a forecast change that much in just twelve hours? The answer is that weather is a complex system that feeds back on itself. In this instance a Canadian high-pressure system moved faster than expected because of influences from other Northern Hemisphere frontal systems. The high-pressure system deflected the snow-producing, low-pressure system farther to the south. I doubt a butterfly in China is the cause of this change, since it is winter there as well, but here in 2004 Lorenz's sense that consistently accurate weather prediction would never be possible is holding true.

In Lorenz's day, complex systems like the flow of commuter traffic and weather were called chaotic because it wasn't possible to predict what they would be doing at an exact point in time—as opposed to a linear system, which is predictable. Since prediction wasn't possible, commutes and weather were seen as chaotic, messy things. I agree that particular kinds of weather and commutes can be quite messy, but in fact such systems are not chaotic at all. If we examine the patterns generated by commuters or weather over larger scales of time, like a year or a decade, then the behavior of the system becomes quite conservative and predictable. It's true that we can't accurately predict the weather one week from now, but based on many years of data we can confidently assume that January will be the coldest month of the year in Vermont, that we won't get low-elevation snows in July, and that November will be our cloudiest month. The patterns consistently repeat themselves. So these systems are not chaotic, it's just that we can't predict exactly what they will be doing at any particular point in time. Because of their conservative, long-term behavior, scientists no longer call these systems chaotic or the study of them chaos theory. *Nonlinear systems* are now called complex systems and the study of them complex systems science.

Attributes of Complex Systems

Because of feedback, complex systems share a number of attributes not observed in linear systems. These attributes will be instructive in our examination of progress, since all socio-economic systems are complex. The attributes I will focus on are: *emergent properties*, *self-organization*, *nestedness*, and *bifurcation*.

Since the parts of a complex system can interact in numerous ways, researchers in this arena quickly realized that it was much more productive to study the interactions between the parts as well as the pattern (or system behavior) that emerged from those interactions—rather than the parts alone. This represents a return to the process and pattern of Aristotle. What was also realized is that the system behavior or pattern was far greater than the sum of a system's parts. Complex systems generated emergent properties—things that couldn't be predicted by just examining the parts.

A trip to the savannahs of Kenya brings people into direct contact with one of the most impressive animals on the planet. It's not the elephant, giraffe, or lion: rather, it's the African termite. These termites build huge mounds, some up to more than twenty feet in height and as large as a small house. Big mounds house colonies that number in the tens of millions of termites. By examining individual termites, would it be possible to predict they could create such massive structures or that they can maintain an internal mound temperature that varies by only a few degrees?

Activities within a termite colony are controlled by the queen. The queen is trapped—with her king—in an underground nuptial chamber. Besides controlling the colony's behavior, the queen's other role is to produce about 100,000 eggs a day to keep the colony well stocked with workers and warriors. To create this kind of egg production the queen's body grows to immense size in comparison to her workers. While a worker is the size of a small ant, the queen is as fat as a person's thumb and about a half-foot long. This leaves the queen unable to move or even care for herself. She has to rely on her workers for everything, including feeding.

Termites communicate via the chemistry of their saliva. Whenever two termites meet, they "kiss" and exchange saliva, transmitting chemical messages. If the mound gets too warm, the chemistry of workers' saliva changes. As workers meet, the chemical message is passed throughout the mound. Eventually the message is passed to the queen during feeding by a worker. Along with eggs, the queen also produces a continuous, chemical-rich secretion that exudes from pores all along her abdomen. Workers constantly suck up these secretions along with their chemical messages and pass them along through the colony. When the queen picks up the message through her feeding that the mound is getting too hot, the chemistry of her secretions changes. Workers tending the queen pick up the new chemical message and carry it out into the colony. Workers in the colony who receive this chemical message stop what they are doing and make their way far underground to the water table, where they fill themselves. They then climb back up into the mound and paint its walls with water. Evaporation of the water lowers the mound's temperature, and when the correct level is reached, chemical messages come back to the queen and her water-gathering message is turned off.[5] The mound, the maintenance of its temperature, and the chemical communication pathway are all emergent properties of the termite colony that couldn't be predicted from examining individual termites. A focus on individual parts—the termites—in a reductionistic, linear approach would completely miss the large-scale behavior of the colony.

In a way the termite colony is a superorganism, with warriors functioning as the immune system, workers as the nervous system and musculature, and the queen as the brain and reproductive organs. Like the termite colony, all organisms are complex systems—something lost to Descartes. As such they have emergent properties. Through the study of neurons and neural pathways in the brain, who could predict emotions such as love and sorrow, or the propensity to create music and poetry? An examination of the parts could never surmise these attributes because they are emergent properties that arise from the complex interactions within the brain. In a complex system the whole is always greater than the sum of the parts.

Many complex systems also function in a similar way to Aristotle's self-completing entelechy. But the term used today is self-organizing. These are systems that take in energy and use it to increase their level of complexity through time. All natural complex systems that grow (and many human-generated ones) function in this way. Multicellular organisms start life as a single cell and, as they develop, increase that cell to millions of cells representing a variety of specialized cell types. A clearcut forest is left in a simplified state. In time it grows back to a forest with complex structure and a wide variety of organisms. Life on Earth started as single-celled organisms restricted to marine environments. Now even the complexity of life on land is astounding. Hurricanes start as small tropical depressions that feed off the energy of warm ocean waters and grow and increase their internal complexity. The Internet didn't exist prior to 1990. Its level of complexity today is mind-boggling. All these systems are engaged in self-organization and as such grow and increase their complexity through time. A linear system by comparison is static and unable to self-organize.

Complex systems also tend to be nested, one within another, and are separated by *fuzzy boundaries*. These are boundaries that allow for the flow of energy, materials, and information between larger- and smaller-scale systems, but maintain each system's integrity. Biological and political systems demonstrate these attributes well.

Within an *ecosystem* are individual organisms that take in energy, matter, and information from the ecosystem. For humans our digestive system, lungs, skin, and sensory organs form our fuzzy boundary. Within each of us lie our cells, which have a membrane to function as their fuzzy boundary to regulate the movement of molecules into and out of the cell. Within each cell are organelles, such as mitochondria, that also separate themselves from the rest of the cell by a membranous fuzzy boundary. All of these represent a series of nested, complex, biological systems—mitochondria in the cell, the cell in the organism, the organism in the ecosystem, the ecosystem in the *biosphere*.

We can see the same nestedness in political entities. Towns and cities are nested in states, which comprise the United States of America—one of almost 200 countries on this planet. All of these different levels of political organization have spatial boundaries that give them integrity and laws that govern how those boundaries operate. Because materials, energy, and information cross into and out of towns, states, and countries, they too have fuzzy boundaries like nested biological systems. As we will explore in upcoming chapters, self-organization and nestedness will be critical in our examination of progress, as will be the process of bifurcation. However, before delving into bifurcation, let's examine how an *old-growth forest* displays emergent properties, self-organization, and nestedness.

INTO THE FOREST

A forty-minute drive from my home in Westminster, Vermont, takes me to the southwestern corner of New Hampshire and to Pisgah State Park. At 13,000 acres, Pisgah is the second-largest state park in New England. Only Baxter State Park in northern Maine is bigger. Pisgah is a jewel of wilderness. Unlike other state parks it is completely undeveloped, with the exception of trailhead parking lots, and much of the western half of the park was never cleared for agriculture during the nineteenth century. As such, Pisgah offers the largest expanse of intact low-elevation forest in the region. In the core of its western side, one can still find pockets of old-growth forest. During the last few years I have come across four such pockets, the biggest covering almost 100 acres on the southeastern side of North Round Pond. We will be visiting this largest stand of old growth in each chapter to explore how it displays the scientific principles discussed in this book.

As I approach the old-growth from the east, I pass through a forest, dominated by sixty-five-year-old hardwoods, that was spawned after the devastating winds of the 1938 hurricane. This southeastern-facing slope took the full brunt of the storm—the entire forest was leveled. Since the hurricane, the young hardwood forest has increased its level of self-organization as it has grown in stature. Up ahead I can see the crest of the ridge. When I reach it and start to descend the northwest-facing slope, I will enter a wind-protected forest dominated by 350-year-old hemlocks.

The hemlocks appear as stately as some of the grand trees that exist in the forests of the Pacific Northwest—their coarse, rusty-red bark graced by a chartreuse-colored crustose lichen. I have only seen this lichen growing on hemlocks of more than 250 years, so its presence is a testament to their age. The forest is anything but neat and tidy. Dead chestnut snags are lodged in the hemlocks; the rotting remains of snapped-off trees litter the ground. But hidden in what appears to be disarray is in fact a very complex and highly ordered system.

The energy that the trees capture from sunlight is partitioned into myriad pathways, each supporting, particularly below ground, innumerable organisms. Since it is not a linear system, we can't predict how each bundle of energy will move through the forest, but we can predict that for every bundle captured by the trees an equal amount will be released as heat from the forest. In this way the forest is a stable system, anything but chaotic.

On the trunk of a nearby white ash is some lungwort—a large-bodied lichen that looks like a lettuce leaf tacked onto the bark. Lungwort isn't a plant, but rather an association between a green alga and a fungus. In this case the photosynthetic algae are nested within the thallus of the fungus. Each lungwort is nested upon its own tree. All the trees are nested in the old-growth forest. This is just one example of the overlapping nestedness that occurs in this forest. If we could see into the soil, the layers of nestedness would be even more complex. It has been estimated that a teaspoon of soil contains billions of organisms, all connected in a dizzying array of interactions. The trees, the soil organisms, and all the other plants, animals, and fungi in this forest, while striving for their own existence, create a system that supports all. Each organism has developed numerous tightly knit relationships with its neighbors. Through these interactions each species has not only become more specialized to coexist with its neighbors, but to service them as well. Ecologists call a forest like this one an ecological community. It is an apt term since all the species in this old-growth forest help each other thrive. This is a clear statement of the system's high level of self-organization.

At some time in the past, it's quite likely that this site didn't support an old-growth forest. Maybe a winter gale leveled a previously existing forest more than a thousand years ago—not unlike what the 1938 hurricane did to the forest on the other side of the ridge. Such an event would have disrupted the high level of self-organization in that community. However, through time the new maturing forest would steadily increase its complexity and self-organization, eventually reaching its present state. Just like the growth of an embryo to a fetus, to a newborn, to a child, to an adult, forests through time increase their complexity and self-organization. They also develop emergent properties.

One of these properties is that there is no such thing as waste in this forest. Every discarded bit of biomass, secretion, excretion, necrotic mass, and even the slime trails of the slugs that forage algae on the trunks of the beech trees are a resource for other organisms. Everything is consumed and recycled in this forest. There would be no way to predict that by examining each organism; it is a property of the entire system and a testament to its very high level of self-organization. A forest like this can offer an Edenic blur of beauty and peace, but a deeper look opens an incredibly intricate complex system that sustains not only itself but all that live within it. Nature can teach us a wealth of wisdom if we just pay attention.

Bifurcation

There are two kinds of feedback that can occur in a complex system—positive and negative. Although positive and negative are often equated to good and bad, in the context of complex systems science there is no such relationship. *Negative feedback* maintains the status quo of a system's behavior. The termite colony and mound represent a system with negative feedback that keeps the mound's temperature at a consistent level. When the mound heats up, the queen sends out a message for the workers to get water. As the mound cools down through evaporation, the queen is alerted through a chemical message. The shutting off of the queen's message to get water is the negative feedback in this system that keeps the mound's temperature at a set level.

In a system with positive feedback, the feedback amplifies the system's behavior in a directional, accumulative way. Today sea ice in the Arctic Ocean is thinning due to a warming polar climate. If the current rate of sea ice thinning continues, within 20 years a large portion of the Artic Ocean will become open water during the summer season. The open ocean will absorb sunlight that otherwise would be reflected out to space off the sea ice, thus generating further warming. This represents positive feedback in the polar climatic system. With sustained positive feedback the impacts eventually may build up to such a degree as to throw the system into a totally new mode of behavior. The point at which a complex system jumps into a new behavioral pattern is known as a bifurcation event.

We can see bifurcation events all around us. An earthquake, where pressure in the Earth's crust consistently builds for years or even decades and eventually results in a powerful temblor; or a dramatic drop in the stock market, where stocks consistently being overvalued for years are dumped all at once; or a revolution, where years of oppression eventually result in citizens overthrowing their political system—all happen quickly, often without warning, the result of positive feedback within the system.

Bifurcation may be the most challenging concept related to complex systems science because it presents a new paradigmatic view of change. Since the time of Charles Darwin Western culture has viewed change as a slow, gradual, accumulative phenomenon. But due to bifurcation points in a complex system, large-scale change happens dramatically fast, not gradually. Although the positive feedback leading up to bifurcation may be gradual, the change in system behavior is abrupt—the proverbial straw that broke the camel's back. The positive feedback is the continued loading of the camel. As this is happening everything seems fine and there is no indication that the camel is going to have a problem until that last straw is added and the bifurcation occurs resulting in a broken back.

This paradigm of rapid change brought forth in complex systems science is actually not new. Complex systems theory has revived an eighteenth-century view of change. Before Darwin's theory of evolution was published, the reigning paradigm in Western science was that large-scale change was abrupt. The formation of mountains, oceans, and new species was due to direct and cataclysmic intervention by God. But in the late eighteenth century James Hutton brought forth the geologic theory of *uniformitarianism*—the idea that all geological features could be explained through slow, accumulative change. Over long periods of time sediments could pile up under the ocean and then be lifted slowly to form mountains—gradual, accumulative change.

Hutton's work dramatically influenced the geologist Charles Lyell, who was a close friend of Darwin and had a profound impact on the young naturalist's thinking. Darwin incorporated this view of slow, accumulative change into his theory of the evolution of species. His one problem was that the fossil record didn't support his theory of gradualism. Examples of fossil sequences that showed gradual, accumulated change were rare, while abrupt changes of fossils found in rock strata were common. Darwin claimed that the fossil record hadn't been studied completely and that when it was, it would validate his theory of gradualism.

For more than a century paleontologists examined the fossil record in rock strata around the world and usually found long periods of little change in fossils followed by abrupt changes in the strata. In 1972 Niles Eldredge and Stephen J. Gould came up with a new theory of evolution called *Punctuated Equilibria.* In this evolutionary theory, the norm for species is stasis or equilibrium, which means that they show little evolutionary change over long periods of time. This would be equivalent to a complex system's conservative approach to maintaining its system behavior through time via negative feedback. Eldredge and Gould theorized that the long period of stasis is then followed by dramatic punctuated change—a bifurcation event known as speciation. A debate about whether new species evolve through gradual processes or quick, punctuated events is currently active in the biological community, but based on our understanding of how complex systems behave, Eldredge and Gould's theory makes sense, since species are complex systems.

Another critical aspect regarding bifurcation events is that their timing often can't be predicted. In fact, they sometimes come as a complete surprise because the positive feedback is masked by the system's status quo until the point of bifurcation is reached. A wonderful example of this is the fall of the Berlin Wall. No one predicted this event a year, a month, or even a week before it happened—the CIA was caught completely off guard. From all appearances Communism was well in place. But positive feedback had been at work in the system for years, and on that day in 1989 a bifurcation was reached and the whole political system changed overnight.

The Linear Paradigm and Western Culture

For the past four centuries a linear, reductionistic paradigm has not only structured Western science but also the culture's view of the world. Our educational system has played a strong role in developing and supporting this paradigm. All of my science classes in high school and college were based on a linear systems approach, even though we were often studying complex systems. Not only that, the entire way the educational system was structured was based on the linear paradigm. Knowledge was divided into parts—science, philosophy, history, math—when in fact all these approaches are tightly connected. During my education I was never exposed to an interdisciplinary approach to knowledge. Knowledge was divided into its parts, and then the parts were divided further. Science was broken down into the biological and physical sciences. Then each branch was reduced further and further into even more specialized courses—reductionism. Most of the courses also had a linear approach to evaluation that focused on the parts through objective tests. To prepare for these tests I had to memorize thousands of individual facts. It was rare to take a test that asked me to synthesize the material in ways that examined my understanding of pattern or process.

Based on this educational approach, it shouldn't be surprising that our entire cultural outlook is focused through the lens of linear-systems thinking even though the bulk of the systems we see and interact with don't function in this way. As such we are deceived into a sense that we can control things like nature, the economy, or social problems by tinkering with parts, when in fact we can't. A look at the 2003 foreign policy on terrorism is a prime example.

Since international relations form a complex system, a sound approach to solving world terrorism would be to look at the large-scale system's behavior that has given rise to it. What has been the nature of the positive feedback that has created and feeds this fearful state of affairs? This question should be the focus of our foreign policy if we are to solve this problem. Yet our current foreign policy is based on a linear approach in which we focus on the parts—Bin Laden, Al Qaeda. Our leaders think we can control terrorism by simply taking out the terrorists. Although going after terrorists should be part of our policy, by itself this approach—focused solely on the current individuals and organizations who commit crimes of terror—will never succeed in ridding the world of terrorism. It will only maintain the "war on terror" indefinitely as ever-new terrorists replace the old. If we hope to win this war then we need to remove the positive feedback that breeds terrorism. The problem is that our leaders haven't been trained to think from a complex system's perspective. They are still enmeshed in a paradigm where a system, like international relations, can be controlled unilaterally by tinkering with its parts.

The huge growth in the use of pharmaceuticals is another example of a linear approach to problem solving. Rather than identifying the underlying causes of disease and focusing our efforts there, most of the funding and research go into figuring out how to fix a disease after it has taken hold. Again this represents a linear approach focused on manipulating the parts—symptoms of the disease—as a means to control the disease itself, instead of adjusting the system's initial conditions, which are often environmental, to keep the disease from occurring in the first place.

Predictability and control lie at the heart of our reigning notions of progress. Our leaders believe they can control the future by constantly adjusting the parts. Technological advances are touted as the means to control one day those things that we can't control right now, allowing progress to continue. Yet the systems relating to progress—social, political, economic, and environmental—are complex, and as such can't be controlled in this manner. In fact the more we attempt to control them, as the fall of Communism points out, the more we tend to force the system, through positive feedback, into an entirely new mode of operation. Control is a reality in a linear system, but in a complex one, it's simply a myth.

Notes

1. Rachel Carson, 1962. *Silent Spring* (Greenwich, Conn.: Fawcett Publications, Inc.), 261.

2. J. Gleick, 1987. *Chaos: Making a New Science* (New York: Penguin), 12.

3. Ibid., 16

4. Fritjof Capra, 1982. *The Turning Point: Science, Society, and the Rising Culture* (New York: Simon and Schuster), 61.

5. Joan Root and Alan Root, 1978. *Castles of Clay* (Briarcliff Manor, N.Y.: Benchmark Films, Inc.).

2

THE MYTH OF GROWTH

Limits and Sustainability

> "Exponential growth does not continue forever. Growth of population and industrialization will stop. If man does not take conscious action to limit population and capital investment, the forces inherent in the natural and social system will rise high enough to limit growth. The question is only a matter of when and how growth will cease, not whether it will cease." —Jay Forrester[1]

Carrying Capacity and Limits to Growth

On May 11, 1943, American forces stormed Attu—the island that anchors the western end of Alaska's Aleutian archipelago. At the time Attu was held by more than 2,400 Japanese troops. The next eighteen days witnessed one of the toughest battles in the Pacific Campaign of World War II. By May 29 only twenty-eight Japanese soldiers remained alive. The cost in American casualties was heavy as well, with seventy-one men being killed or injured for every 100 Japanese. Only the battle for Iwo Jima resulted in proportionally higher losses for the Americans.[2]

By August 24, 1943, the Japanese had been forced out of all the Aleutian Islands, but as part of the war department's plans against future Japanese activity in the region the U.S. Coast

Guard put a LORAN (long-range aid to navigation) station on St. Matthew Island to the north of the Aleutians. St. Matthew—an isolated island in the middle of the Bering Sea—was manned by a contingent of nineteen men. As a means to supply fresh meat to the operators of the station, the Coast Guard introduced twenty-nine reindeer to the island in August 1944. With the surrender of the Japanese the following year, the St. Matthew station was closed; the reindeer herd was forgotten.

Twelve years later, U.S. Fish and Wildlife biologist David Klein was sent out to St. Matthew to see what had become of the reindeer. With the help of associates, he inventoried the entire 130-square-mile island and counted 1,350 healthy animals—a forty-seven fold increase in the size of the herd in little more than a decade.[3] The reindeer's dramatic population growth was the result of two things: good habitat in the form of dense stands of reindeer lichen—a staple for this species—and a lack of any predators.

Klein's next visit to the island occurred in 1963. Although the population growth rate had slowed during this interval, Klein now counted 6,000 animals. This time the reindeer were noticeably malnourished, having lost an average of 40 percent of their body weight since his last visit; their forage in reindeer lichen was dramatically degraded.[4] The following winter was harsh, even by regional standards, but Klein wasn't able to return until the summer of 1966. That summer he found that the reindeer population had suffered a precipitous decline. Skeletal remains were everywhere, and his inventory found only forty-two remaining animals—all females with the exception of one deformed male.[5] By this point the herd was no longer able to reproduce, and the entire reindeer population became extinct sometime in the 1980s. The

St. Matthew's reindeer herd is a potent example of what can happen to a population that exceeds its *carrying capacity*.

Carrying capacity is an ecological concept that is defined as the maximal population size that an ecosystem can support without being degraded in some fashion. Once a population of organisms overshoots its carrying capacity, the ecosystem that supports the population becomes impaired, which in turn has negative consequences for the population that has grown too large. What results is a precipitous population decline. The St. Matthew herd is an unusual example since it is very rare for even a dramatic population die-off brought on by overshoot to result in extinction. The catastrophic reduction of the reindeer's food supply, caused by overgrazing, resulted in the degradation of the population's ecosystem. With so many animals entering the winter of 1964 in a malnourished condition, only a handful were able to survive. In this way the limited forage generated negative feedback within the ecosystem, and negative feedback led to the eventual demise of the reindeer.

Every population of organisms in every ecosystem on this planet has a carrying capacity. If they exceed it, their system will be degraded, creating negative feedback that will eventually cause a decline in the organism's population size. In this way the carrying capacity sets the *limits to growth* for all populations.

Another Arctic species that regularly taunts limits to growth is the lemming, whose populations exceed their carrying capacity roughly every four years. Some of you have probably heard of lemmings throwing themselves from cliffs when their population growth has reached its zenith. The notion that lemmings kill themselves in such a fashion is actually a cultural myth that was birthed by less than a minute of footage from one of Walt Disney's early nature documentaries, *The Living Arctic.*

When lemmings exceed their carrying capacity, both a lack of suitable forage and an increase in stress—due to frequent encounters with members of their own species—eventually cause individuals to leave their territories in an effort to seek less populated locales. Lemmings migrating randomly across the landscape displace ever more lemmings from their territories. In their agitated, migratory state they suffer high mortality rates from increased predation and malnourishment brought on by reduced forage.[6] But Walt Disney knew that random migration and death of lemmings would not serve as very dramatic footage. As a means to add drama to *The Living Arctic*, he orchestrated a more memorable ending for the lemmings. His crew gathered up hundreds of these animals and transported them to an oceanside cliff. With cameras running the crew released the lemmings and stampeded them off the cliff. To this day, most Americans familiar with lemmings believe that somewhere out there exists "Lemming Cliff" where these animals regularly hurl themselves into the sea. Although lemmings do suffer dramatic population decline following overshoot, intentional suicide is not one of the reasons for the decline.

When I was in high school I witnessed the impacts of another example of a population exceeding its carrying capacity. At the time my family lived in a somewhat wooded suburban neighborhood with houses on two-acre lots. In the center of the neighborhood was a small pond, roughly five acres in size, surrounded by about a dozen homes. The pond had been stocked originally with bass, providing some decent recreational fishing for children as well as adults. As homes around the pond changed hands, new homeowners started removing the woodland portions of their properties and converting those areas into lawns that ran right up to the water's edge. The pond and its associated lawns soon attracted Canada geese that started to over-winter since many families fed the geese leftover pieces of bread and even cracked corn. It wasn't long before the nature of the pond changed visibly from all the nutrients introduced to it via the runoff of lawn fertilizers and the droppings of geese.

During the summer the pond became choked with billowing clumps of filamentous green algae—some as large as cars. One summer when the bloom was particularly bold, within about a week's time all the algae died and floated to the surface, creating a fetid brown mat in which the carcasses of dead bass and sunfish were commonly trapped. The algae had exceeded their carrying capacity, not only to their own demise, but also to that of the fish and possibly other organisms. When the population of algae shot past its carrying capacity, its demand for oxygen during warm summer nights couldn't be satisfied. The dissolved-oxygen levels in the water at night would plummet, degrading the entire pond ecosystem and eventually resulting in the death of both algae and fish.

The Biosphere and Dynamic Equilibrium

Limits to growth occur whenever one system is nested within another, like a population of algae nested within a pond. Since the pond has finite resources, in this case dissolved oxygen, once a population digs deep into those resources negative feedback occurs and results in the population's decline. Although most of the scientific research on limits to growth has been conducted on populations within ecosystems, all of the natural complex systems that we can observe on this planet—biological, ecological, geological, meteorological—also function under the overarching law of limits to growth. This even holds true for the largest biological system known—the biosphere.

The biosphere is comprised of all the ecosystems that cloak the Earth. Current scientific evidence suggests that living organisms have been on this planet for at least 3.5 billion years and photosynthetic organisms—those that can convert sunlight into usable energy—have been around for at least the last 3 billion years. Since that time organisms on the Earth have captured more energy through *photosynthesis* than they have released as heat through their metabolism or *cellular respiration.*

Photosynthesis is a process where water and carbon dioxide are combined using captured solar energy to build carbohydrates—energy storage molecules such as sugar or starch. In these carbohydrate molecules energy is stored wherever a hydrogen atom is bonded onto a carbon atom. Along with the production of carbohydrates, oxygen is released as a byproduct of photosynthesis. Cellular respiration is photosynthesis exactly in reverse. Oxygen is used by cells in order to break down carbohydrates and retrieve the energy stored within these molecules. Once used by the organism, the retrieved energy is lost as heat. Along with heat the two other byproducts of respiration are water and carbon dioxide.

Photosynthesis: carbon dioxide + water + solar energy > carbohydrates + oxygen
Respiration: oxygen + carbohydrates > heat energy + water + carbon dioxide

If a plant conducts more photosynthesis than respiration, it grows. The growth can be measured as the increased *biomass*

—organic matter—of the plant from the carbohydrates stored from photosynthesis. If photosynthesis and respiration are equal in a plant, growth stops and *dynamic equilibrium* results; in this state the solar energy absorbed by the plant is equal to its energy released as heat.

Scientific evidence suggests that the early atmosphere of the Earth prior to photosynthetic organisms was devoid of oxygen gas. As photosynthetic life spread across the globe, oxygen started to build up in the atmosphere. Increasing oxygen is synonymous with growth because it means that photosynthesis and the production of biomass is taking place at a higher rate than respiration, a process that consumes oxygen. If both of these processes had occurred at an equal rate since life first established on Earth, today the planet would be devoid not only of an oxygen-rich atmosphere but also of all living and dead organic matter, since each represents stored biomass energy which has to result from historically greater rates of photosynthesis.

Atmospheric oxygen reached its current 21 percent concentration more than a quarter billion years ago. Since that time it has been quite stable, signifying that biospheric photosynthesis and respiration have equalized, creating a dynamic equilibrium where the rate of solar energy captured by life on Earth is equal to the rate it is released as heat through the metabolism of all its biota. This heat eventually drifts out to space where it is dissipated into the Universe. Like reindeer, lemmings, and algae, the biosphere honors limits to growth, too.

From the window behind my computer, I look out to what used to be a pasture for dairy cows and has now grown into a young white pine woodland. Below the pines, red maple and black birch saplings have started to invade the forest understory. From the older forest behind this pine woodland I'd guess that in time, after the maple and birch gain dominance, that they will eventually be displaced by eastern hemlock. Although the chances are extremely slight, with protection and luck this patch of forest may avoid future disturbances such as logging, ice storms, pathogens, or blowdowns and reach old-growth status with hemlock trees more than 300 years of age. If this happens the forest will experience more than 400 years of *successional* change—pasture to pine to mixed hardwoods to hemlock to old-growth forest.

INTO THE FOREST

Although it is unlikely that the abandoned pasture on my property will reach old-growth status, here in Pisgah I experience the real thing. Through the centuries this forest experienced continuous growth as it accrued ever more biomass. That biomass is now visible in huge, widely spaced hemlocks whose four-foot diameter trunks rise sixty feet before spreading branches. It is visible in the moss-covered, moldering, downed trunks of trees that lived a long, full life. It is visible in the dark humus of the soil that is filled with unseen organisms. Everywhere I look I see stored solar energy. It is truly mind-boggling, but everything I see, both living and dead, is mostly composed of just water and carbon dioxide stitched into new forms by sunlight. In this forest, as in any organism or ecosystem, very simple beginnings have brought forth a complex majesty.

Photosynthesis has well surpassed all of the biota's respiration over many centuries in this old-growth, but today, like yin and yang, these two opposing dynamics are balanced. I can't see this balance; on any given day or month photosynthesis may outpace respiration or vice versa. However, over a larger time frame such as a decade, all of the solar energy absorbed in photosynthesis will be pretty much equal to that released as heat through the cellular respiration of all the forest's organisms. In terms of energy flowing into and out of this system, the old-growth is a dynamic equilibrium. It will maintain this state until at some point it experiences a major disturbance, for example a wind or ice storm. Until then its hemlocks won't grow bigger than the largest trees I experience today. The number of downed trees will remain much the same as young trees replace the old. The humus layer won't grow thicker. Each year as hemlock needles and leaves of beech and maple fall to the forest floor, an equal amount of organic material in the soil will be broken down to carbon dioxide, water, and heat to be released from the forest. Like a teenager entering adulthood, growth has stopped for this hemlock forest. This is not just true for forests; all ecosystems eventually reach a dynamic equilibrium where the energy entering the system is equal to the energy being released as heat.

Dynamic Equilibrium of Other Natural Systems

Although we have no memory of it, each of us was once a single, microscopic cell. This is certainly astounding to contemplate. As we developed from single cell to fetus to newborn to adult, we took in more energy than we released through respiration. As a result we stored that energy in our increasing growth as that one cell grew into more than thirty trillion.

Thirty trillion is a number that is hard to comprehend, so an analogy may help. Imagine a beach backed by dunes. Between the dunes and where the ocean's waves break at high tide is a distance of 100 feet, and the depth of the dry sand where the beach abuts the dunes is three feet, so the average depth of dry sand on this beach is 1.5 feet. You walk onto the beach and scoop up a large handful of sand. This equals about

100,000 grains. How much of this beach is required to hold thirty trillion grains of dry sand? To get this many grains of sand we would need a length of beach equal to almost a mile. All of those grains of sand on that mile length of beach are equivalent to the number of cells in your body!

When we reach adulthood we stop increasing the number and size of the cells in our bodies. The only exceptions are fat and muscle cells that can increase in size in an adult. Yet for an adult with a healthy lifestyle, growth stops and like an old-growth forest the body reaches a dynamic equilibrium in which the number of calories assimilated from food on a daily basis equals the number released through body heat. Like us, all other organisms honor limits to growth at adulthood. This is a trait that has been selected by nature. Any individual that continued to grow and needed ever increasing food resources after it reached adulthood would have more and more difficulty meeting its energy needs compared to individuals who stopped growing.

Limits to growth is an inherent law that governs all organisms, populations, ecosystems, and even the biosphere—systems that are all nested one within the other. But not only do biological systems honor this law, it operates in the Earth's physical systems too.

On September 4, 1938, an upper-air depression formed in the south-central region of the Sahara Desert. Little could anyone have known that in seventeen days this meteorological phenomenon far off in Africa would generate the most powerful hurricane to strike New England in its recorded history—the Great Hurricane of '38. As the storm migrated over the middle Atlantic, it absorbed energy from the ocean's warm waters. It used this energy, like biological entities, to grow and become more organized. By September 16, located to the northeast of Puerto Rico, the tropical depression had grown and become energized enough to reach hurricane status. For the next few days it moved northward, strengthening to become a strong Category 3 hurricane. But it then stabilized on the evening of September 20.[7] This storm, like all hurricanes and typhoons, eventually reached a state where it didn't become more powerful. Nested in the global meteorological system, these cyclones honor limits to growth since the energy resources that support them are finite.

The same is true for glaciers, mountains, mountain ranges, and any complex physical system that exhibits growth. All eventually reach a state of dynamic equilibrium where growth stops. In fact, try to think of a single example of a biological or physical system that experiences ever-increasing growth.

Neoclassical Economic Orthodoxy and Growth

This then raises the question: if everything that we observe in the world around us honors limits to growth as a means to sustain itself, why is the underlying foundation for our current paradigm of progress ever-increasing growth? The answer lies in a body of economic theory that not only has no grounding in, but is actually divorced from, the scientific laws that govern the Universe. In a very real sense our reigning neoclassical economic orthodoxy has been developed in an artificial world where resources are infinite, and waste, including garbage, pollution, toxins, and environmental degradation, don't exist, and where our socioeconomic system functions in a void rather than being nested within the biosphere.

Classical economic theory was developed in the eighteenth century prior to the Industrial Revolution. At that time resources seemed infinite and the kinds of waste generated by industrialization had not yet been experienced. Adam Smith's influential work, *The Wealth of Nations*, was published in 1776, a good two decades before the onset of industrial culture. Since resources appeared unlimited and industrial waste was nonexistent, it is understandable that classical economic theory was solely focused on human commerce, wealth, and market dynamics—excluding any interface with the natural world. Economics was birthed in a void. To this day, classical theory reigns, even though there are a large number of economists who attack the governing economic theory as being flawed.

The first authoritative voice to state that our current economic orthodoxy was seriously flawed since it didn't incorporate critical scientific laws was Nicholas Georgescu-Roegen. In his 1971 book *The Entropy Law and the Economic Process*, Georgescu-Roegen does a masterful job of pointing out how reigning economic theory doesn't incorporate the second law of thermodynamics and the law's associated implications for ever-expanding energy usage. This work was followed in 1972 by the powerful book *The Limits to Growth* by Donella Meadows and associates—all accomplished systems scientists. In this book they point out how our economic system is not sustainable and would run into serious consequences in the twenty-first century. In *Steady-State Economics*, published in 1977, Herman Daly builds on the work of Georgescu-Roegen and points out the inherent flaws in an economic theory based on continual growth. In this book Daly develops an alternative economic system that operates as a dynamic equilibrium. More recently William Miernyk in *The Illusions of Conventional Economics* and Steve Keen in *Debunking Economics: The Naked Emperor of the Social Sciences* criticize reigning current economic theory as being scientifically unfounded.

Kenneth Boulding, former president of the American Economic Association, once stated, "Only madmen and economists believe in perpetual exponential growth."[8] Yet these voices are rarely heard in our current political climate and are certainly excluded from the development of economic policy at both the national and global level.

The governing concept on which our current economic paradigm of continuous growth is based is the *principle of unlimited substitutability.* This economic principle states that resources are, figuratively speaking, unlimited because as we exhaust one resource we will replace it with another, so growth will never cease. We can continue to do this as long as we develop new technologies and increased access to new energy resources. It is precisely these new technologies and increased energy usage that need to be examined more fully since they create a nexus between economic theory and scientific law. Although orthodox practitioners like to claim that their economic theories are scientific, no amount of mathematical models or statistical equations can label something as science if it intentionally ignores foundational scientific laws such as limits to growth or the *second law of thermodynamics.* But before we get to the second law of thermodynamics and its relationship to energy consumption, we should take a closer look at technology—the other half of the questionable foundation upon which the principle of unlimited substitutability rests.

The Cost of Technology

To an orthodox neoclassical economist, technology is a free ride where we can get something—increased access to resources—for nothing. Technology offers only benefits—

cheaper energy, greater food production, greater longevity, faster travel, expanded communication networks. Yet technology is not benign. All new technologies have some costs. The negative costs of technology are probably best described in chapter nine of Barry Commoner's classic, *The Closing Circle.*

In this chapter, titled *The Technological Flaw*, Commoner examines the rise of pollution in the United States during the 25-year period from 1946 to 1971. During that quarter of a century the U.S. population increased 42 percent, the Gross National Product rose 126 percent, and consumption of food, clothing, and shelter increased by 40 to 50 percent—directly linked to the increase in population. Yet these levels of growth alone could not explain the 200 to 2,000 percent increases in pollution. The difference was largely due to new technologies that brought about the replacement of pre–World War II materials with new synthetic ones.[9]

Just a few of the product changes Americans witnessed during those twenty-five years were organic fertilizers being replaced with inorganic fertilizers; soaps being replaced with synthetic detergents; glass, metal, and rubber being replaced with plastics; and not to mention the use of a whole new host of synthetic pesticides. All of these new products were derived from petroleum and demanded much higher energy consumption in their production, creating markedly increased rates of air pollution. But that was only part of the problem. The increased use of chlorine gives a fuller picture.

During this period chlorine became an important chemical product. It was needed in the production of many plastics, such as polyvinyl chloride (PVC), and it was utilized in the manufacture of detergents. Since mercury is necessary in the production of chlorine, its use solely for this purpose increased 3,930 percent between 1946 and 1971. This statistic doesn't include the other dramatic jumps in mercury usage, such as a 3,120 percent increase in the production of paints.[10] With the production of these new synthetic materials, mercury pollution dramatically increased. Today there is so much mercury in our environment that a joint advisory from the Food and Drug Administration and Environmental Protection Agency suggests that pregnant women and young children refrain from eating certain species of large fish, such as swordfish and certain species of tuna, and eat no more than two servings of any fish a week,[11] no matter whether it was caught in a river, stream, lake, pond, or even the ocean. This is how ubiquitous mercury pollution has become.

There were also serious pollution costs following the use of these new synthetics. Many of the detergents and new plastics were not biodegradable since they were composed of new synthetic molecules that no organisms had the means to break down. The only way to get rid of the plastic was to bury it or burn it. Because of the presence of chlorine, burning produced a new toxic substance—dioxin. This is just one example of the negative impacts that resulted from a new technology that allowed us to substitute new materials for old ones, and this example shows that the concept of unlimited substitutability has definite drawbacks.

Due to new technologies in medicine, food production, and energy utilization, the global population of humans soared during the past century from 1.5 billion in 1900 to more than six billion in 2000. Along with this population growth came ever-increasing per capita rates of resource and energy consumption, particularly in developed nations. During the decade between 1991 and 2001, Americans increased their collective consumption by 57 percent.[12] Combined, these two trends have pushed humans beyond the carrying capacity of the biosphere. To reiterate, carrying capacity is the maximal population size that can be supported without degrading a population's ecosystem. However, for humans the concept of carrying capacity needs to be adjusted to account for differing cultures that influence use of resources and patterns of consumption. Not all human populations have the same carrying capacity. Yet our ultimate ecosystem is the biosphere, and collectively as a global population we have exceeded its carrying capacity.

All the global environmental problems that we experience today—pollution in all its various forms, extinction of species, depleted fisheries, dwindling supplies of fresh water, deforestation, global climate change—are examples of a degraded biospheric system. How much further can we go with continuous growth before the negative feedback from these environmental problems has dramatic, visible, global consequences? There is not a clear answer to this question, but I think it's very unlikely that we will be able to maintain in the twenty-first century the rates of growth of global resource consumption witnessed during the last century. The exponential growth rates of consumption will eventually generate such a scale of negative feedback from the global system that growth will be halted.

Every example that we have of systems that can sustain themselves and prosper is based on limits to growth. At maturity they all reach a state of dynamic equilibrium. The idea that unlimited economic growth will bring continuous progress has no scientific foundation. In fact science tells us just the opposite—unchecked growth generates its own negative feedback. We need to embrace another model of progress that is compatible with scientific law, a model already developed by economist Herman Daly. But to engage in such an economic system will first demand major changes in institutions, both corporate and governmental, and possibly more importantly in our own cultural values.

Here is a final example of a system that doesn't honor limits to growth that has touched just about everyone's life—cancer. Cancer is a disease, often environmentally induced, in which the genetic instructions within a cell become altered. In a human body, cells replicate only as needed to replace dying cells, thus the number of cells in a body stays pretty much the same through adulthood. Because of genetic alterations in cancer cells, these cells don't stop their replication when they have reached replacement levels, but continue vigorous, unchecked replication. This results in tumors that grow ever bigger, gobbling up the body's resources and eventually displacing other kinds of important cells. If unchecked the cancer will eventually impair the entire body's system to such a degree that death will result.

Is ever-increasing economic growth and associated resource consumption analogous to a biospheric cancer? Neoclassical economists would say that such an analogy is ridiculous. They would contend that new technologies and access to increased energy will allow us to both grow and fix our environmental problems. The track record of economic growth fixing our environmental problems is not very good to date, as we continue to see an increase of environmental degradation at the global level. More importantly increased energy consumption smacks head-on into the second law of thermodynamics—a scientific law that proponents of our reigning paradigm of progress never mention.

Notes

1. Jay Forrester, 1973. *World Dynamics*. (Cambridge, Mass.: Wright-Allen Press), 144.

2. Brian Garfield, 1969. *The Thousand Mile War: WW II in Alaska and the Aleutians* (Garden City, N.Y.: Doubleday), 256.

3. David Klein, 1968. "The Introduction, Increase, and Crash of Reindeer on St. Matthew Island." *Journal of Wildlife Management* vol. 32: 352.

4. Ibid., 352.

5. Ibid., 352.

6. F. Pitelka, 1973. "Cyclic Pattern in Lemming Populations Near Barrow, Alaska." *Alaskan Arctic Tundra: Proceedings of the 25th Anniversary Celebration of the Naval Arctic Research Laboratory*, 199–215.

7. William Minsinger, 1988. *The 1938 Hurricane: A Historical and Pictorial Summary* (Milton, Mass.: Blue Hill Observatory), 9–10.

8. Garrett Hardin, 1993. *Living Within Limits: Ecology, Economics, and Population Taboos* (New York: Oxford University Press), 191.

9. Barry Commoner, 1972. *The Closing Circle: Nature, Man, and Technology* (New York: Bantam), 138.

10. Ibid., 141.

11. Environmental Protection Agency. See <http://www.epa.gov/waterscience/fishadvice/advice.html>.

12. Juliet Schor, 2002. "Too Much Stuff: Consumerism with a Conscience." *Boston Globe*, 8 December 2002, sec. D, p. 12.

3

THE MYTH OF ENERGY

The Second Law of Thermodynamics

"No domain of the global economic activity does greater social, environmental, and political harm than today's dominant energy systems, from source to waste." —John Cavanagh[1]

The Second Law

Before sitting down to write this morning, I helped my wife, Marcia, clean the house. We swept the wooden floors, dusted the tabletops, and cleaned the sinks and toilet bowls in the bathrooms. If we are good, we do this cleaning ritual about once a week, but longer intervals sometimes pass between cleanings. Since we have been married for more than thirty years we have probably cleaned house more than 1,000 times, and I can't imagine how many times we have just picked up clutter. Why is it that our home, not to mention everyone else's, can't stay organized? Why will dust eventually cover any horizontal surface? The answer lies in the second law of thermodynamics—arguably one of the most important laws of science. As Albert Einstein stated regarding this law, "It is the only physical theory of universal content concerning which I am convinced that, within the framework of the applicability of its basic concepts, it will never be overthrown."[2] The implications of the second law of thermodynamics are so profound, and its workings so pervasive, that it should be an essential part of everyone's education. Yet I fear few understand its workings; this seems particularly true for our policy makers. To begin our examination of the second law of thermodynamics, we need to go back almost two centuries.

While studying steam engine efficiency in 1824, the French scientist Sadi Carnot noticed that heat always flowed from hotter to colder bodies. Carnot's observation certainly wasn't new. But he took his observation a step further by showing that Newtonian mechanics couldn't explain such a unidirectional movement. The result of his work launched a new branch of physics known as thermodynamics. By 1865 the German physicist Rudolf Clausius had drafted the first and second laws of this young science.

The *first law of thermodynamics*, also known as the *law of conservation of energy*, simply states that energy can neither be created nor destroyed. This means that the amount of energy in the universe today is exactly what it was thirteen billion years ago, just after the Big Bang. This is a powerful concept, but from a practical standpoint it's the second law of thermodynamics that is the most important to us. The second law, also known as the *law of entropy*, states that although energy can't be created or destroyed, it can be transformed from one form to another.

As I type, some of the electrical energy that runs my computer originally came from the decay of uranium atoms within the Vermont Yankee nuclear power plant. As a uranium atom breaks apart, a minute amount of the mass of its nucleus is transformed into heat energy. The heat energy is used to produce steam. The steam then turns electric turbines, this transformation producing *kinetic energy*, the energy of motion. The kinetic energy of the rotating turbines is then transformed into electrical energy. The electrical energy enters my computer and is transformed into light and the words that appear on my screen, finally being transformed into heat that dissipates into my room.

Throughout all these transformations no new energy has been created and none has been destroyed. But the transformation of energy from one state to another is not the critical aspect of the second law. The critical point is that although energy can be transformed, no transformation is 100 percent efficient. This means that *within the system* where the transformation occurs, some of the energy is lost from that system during the transformation. The energy isn't destroyed; it simply leaves the system in which the transformation takes place. As we shall see, the concept of nested systems becomes important here because energy lost from one system during a transformation simply dissipates into the larger system around it.

A car offers another example of the second law. The intent of a car is to transform potential chemical energy—stored in the bonds of gasoline molecules—into motion. But this transformation is only about 30 percent efficient. During the transformation most of the energy is lost from the car as heat and drifts off into the air. The loss of energy from a system when energy is being transformed may not at first seem important, but its implications are striking because the loss of energy from a system results in *entropy*.

Entropy

Entropy is a process where things naturally move from a state of order toward disorder, like our homes that we continually need to clean. The problem with order and disorder is that these are subjective terms. One person's view of order can be quite different from another's. A more objective way to describe entropy is as a process in which things move from a state of complexity toward simplicity, or from concentration toward diffusion. In the Universe at large, as time advances, things are becoming more diffused. This is most easily grasped by visualizing the concentrated energy in stars constantly diffusing into space as light and heat or the galaxies drifting apart as the Universe expands. To reverse entropy and create concentrations of energy or materials, or to build complexity, means that energy has to be added to a system and stored within it. The logs that I burn in my wood stove were created from simple molecules—water and carbon dioxide—that were concentrated and bonded together with light energy through the process of photosynthesis to create wood, a far more complex material than the simple molecules from which it was created. So whenever energy is stored within a system, it is stored in ways that increase the system's complexity or concentration of materials. However, whenever energy is lost from a system through transformations, the result is entropy—simplification and diffusion.

Entropy shouldn't be a new concept for us since we all have plenty of experience with its workings. In our homes, each time we do something we are transforming energy and as a result creating entropy, as energy stored within the system from previous cleanings is dissipated. As we browse, books originally neatly stacked on shelves come to lie scattered throughout the house. As we turn each page of the newspaper, some of the paper's fibers are released into the air and drift down as a component of dust. We host a potluck dinner for company and all the dishes, glasses, and silverware that were neatly stacked in their respective cabinets come to grace every horizontal surface. All are examples of entropy caused by our energy transformations that take things from an orderly state of concentration to one of diffusion. We don't intentionally create entropy; it just happens—a result of the second law. However, to reverse entropy takes an intentional effort, like the cleaning of our homes—an investment of energy to increase order and concentration.

When Clausius was formulating the second law, all of his work was focused on closed systems. A *closed system* is one in which energy or materials can't enter the system. A nonrechargeable battery is an example of a closed system. Once the battery is made, it can't accept any new energy. The energy within the battery is stored as potential chemical energy in the form of concentrated ions. With each use, as energy dissipates from the battery, the ions diffuse within the battery until it eventually loses all of its potential chemical energy. When the battery was made it was said to be in a nonequilibrium state—one where the ions were concentrated and as such stored potential chemical energy. Through use and loss of energy, increasing entropy causes the battery to reach a *static equilibrium*—complete diffusion and a dead battery. A state of static equilibrium should not be confused with the dynamic equilibrium of an *open system*.

Unlike the battery, complex systems are not closed. Since they have fuzzy boundaries they can take in energy from the larger system in which they are nested. So there are three outcomes for any complex system. If it releases more energy from its transformations than it takes in from the larger system in which it is nested, it is entropic. A cut down tree is a good example of an entropic system. It can't take in any energy through photosynthesis, so with each transformation created by *saprophytes*—decay producing organisms like fungi—energy that was stored in the tree is released as heat. The complex structure of the tree is simplified as it is broken down to carbon dioxide, water, and other small, inorganic molecules that completely diffuse into the air and soil. However, if a complex system takes in more energy than it releases, it is *anti-entropic.* This means it grows and increases its level of complexity through time. From egg to adult, we are all anti-entropic systems and as such take in more energy through food than we pay out as heat. That energy is stored in the increasing complexity of our growing bodies. As adults we are no longer anti-entropic and, as long as we are healthy, function as a dynamic equilibrium. Healthy adults take in the same amount of energy through food as they release as heat through all their metabolic transformations. Again, this is a dynamic equilibrium in which energy is entering and leaving an open system at the same rates, and it's not to be confused with the static equilibrium of a closed system.

All complex systems that consume energy create entropy in their surroundings—the larger system within which they are nested. A fungus growing on a rotting stump is a good example. The anti-entropic fungus grows by extracting energy from the stump. In the process the stump decays and is simplified. Luckily, the forest in which the fungus and other creatures reside takes in as much or more energy through photosynthesis than it gives off in decomposition and metabolism of all its organisms. So any complex system, whether it is entropic, at dynamic equilibrium, or anti-entropic, is simply defined by how much energy the system takes in versus how much energy is released through its various transformations. This means that the rate at which energy is transformed within a system is an important consideration.

Picture two rooms that are absolutely identical. Each is exquisitely appointed with the same fine antique furnishings, delicate porcelain figurines, crystal vases, rare books, and potted plants. Into each room we will place an individual for one hour. In one of the rooms it will be an adult who will read the Sunday edition of the *New York Times*. In the other it will be an unsupervised toddler. Which room will be more entropic after one hour?

In the adult's room the turning of the pages of the newspaper has generated some entropy in the form of dust that has diffused into the air. The toddler's room will be quite a bit more entropic. The striking difference between the two rooms isn't because toddlers are inherently destructive—they just create a lot more energy transformations as they explore the world around them, which results in greater entropy. The only way to reorder the toddler's room to its original state is to add external energy to that system. The energy is derived from people cleaning up the room and electricity to run a vacuum cleaner.

If the second law didn't exist, we would never have to clean up after ourselves. We could also recycle energy, and as a result make perpetual motion machines where no energy would be lost from a system. We could construct a car so that whenever we braked, all the kinetic energy would be converted into electricity and stored in the car's battery. If there were no losses from any of the transformations in the operation of the car, we would only need to charge the car's battery once and then simply recycle the energy. But in reality, this is not possible—energy can't be recycled. Due to friction in the motor, in the brakes, in the tires on the road, and in the atmosphere against the moving car, energy will be continually lost from the car as heat. Because of the second law of thermodynamics, there is no free ride. We will always pay in entropy for any transformation we create. The only way to reverse entropy is to constantly bring in new energy from outside the system to replace the energy lost through transformations.

All the complex systems that we can observe in the natural world, as they grow and develop, take in more energy than they release and are therefore anti-entropic. But at maturity they become dynamic equilibrium systems in which growth stops, and then at some point become entropic, finally ceasing to exist. All organisms do this as they grow, mature, and die. Even ecosystems do this. Imagine a clearcut and all the energy transformations involved in cutting, limbing, and hauling away the logs. The site is left in a highly entropic state, but allowed to grow back to forest. First herbaceous plants cover the ground, to be followed by shrubs. In time trees invade the site, and after a number of decades a young forest is established. If this forest does not experience another disturbance, after many centuries it will become a mature, old-growth system at dynamic equilibrium. To accomplish this, external energy had to be added to the developing forest. That energy was sunlight, captured through the process of photosynthesis and stored in the increasing complexity of the forest's biomass. Up until the time the forest reaches its old-growth stage, it is in an anti-entropic state, storing energy and using it to increase its complexity.

INTO THE FOREST

In the previous chapter we saw that the old-growth stand in Pisgah is a dynamic equilibrium system in which the amount of energy flowing into the forest through photosynthesis equals the amount of heat being released by all the metabolic transformations of the organisms in the forest. I can see within the forest organisms in a variety of energetic states that together maintain the dynamic equilibrium of the entire ecosystem.

In front of me lie the rotting remains of a huge American chestnut. The original tree must have been about four feet in diameter. It was killed by chestnut blight that heavily impacted this region of New Hampshire during the early 1900s. Across the river in Vermont the blight peaked around 1915, so it is likely that this downed tree died about ninety years ago. Chestnut has wood that normally hollows out as it decays, with the outer wood remaining very rot resistant. This specimen displays that pattern well. The sound, outer wood of the downed log—about three inches thick—has split along its top and fallen back on either side. It lies like a huge dugout canoe—seven feet wide to the gunnels—now filled with rich humus from the decay of its interior wood.

The chestnut clearly represents an entropic system. As each bacterium and fungus that lives in the decaying mass of the chestnut draws energy out of the wood, the tree's highly ordered, complex structure is slowly broken down. First the complex molecules within the cells are broken down, then the lignin between the cell walls—resulting in the wood crumbling into brown blocks of material and then into material completely unrecognizable as wood. In another century, nothing of this tree will be left. It will be reduced to the simple molecules of carbon dioxide, water, nitrogen gas, and other simple, inorganic molecules that will disperse into the atmosphere and soil. All the tree's stored energy will be dissipated as heat to space. Yet this entropic process of decay is essential for the health of the entire forest. It recycles the nutrients of a forest to fuel the growth of replacement trees and other organisms. Without decay all the nutrients would be locked into dead tissues; the forest would soon experience nutrient starvation and cease to exist.

Next to the downed chestnut is a young sugar maple sapling about ten feet tall. I can see from the shiny, brown color of its uppermost branch that this tree grew about six inches in height last year. As a growing tree it represents an anti-entropic system. It is taking in more energy through photosynthesis than it is paying out in metabolic heat. That extra energy can be seen as growth. It is quite likely that many of the simple molecules once lodged in the old chestnut will become incorporated into complex molecules of the growing maple. Like many of the big hemlocks around me, this young tree may make it to old-growth status. When it does, the maple will become an equilibrium system in which all the energy it takes in through photosynthesis will be used in the tree's metabolism. At this point its growth will stop.

Within the old growth we have an array of anti-entropic, dynamic equilibrium, and entropic organisms, all connected in the complex flows of energy and the cycling of nutrients that sustain the entire system. This balance can only be maintained if the amount of energy entering the system keeps pace with the energy being lost through metabolic transformations. However, no forest is immune to disturbance, and eventually this old-growth forest will experience a catastrophic disturbance—entropy—and the entire process will start afresh.

Biospheric Entropy

As discussed in the previous chapter, the biosphere was an anti-entropic system up until about a quarter billion years ago when it entered a state of dynamic equilibrium. However, the biospheric story today is quite different. Since the nineteenth century the increasing use of energy by humans, particularly fossil fuels, has pushed the biosphere out of its dynamic equilibrium state into one that is increasingly more entropic. Human activity on this planet is countering trends that have been developing for over 3.5 billion years. For the first time in the Earth's history, a single species is responsible for the entropic degradation of the biosphere by releasing more energy through transformations than is being replaced by global photosynthesis.

Our species, the modern *Homo sapien*, has been around at least 150,000 years based on fossil evidence. By 10,000 years ago it is estimated that the global population of humans was possibly ten million, existing in small hunter-gatherer clan groups. Ten thousand years ago also marks the beginning of our present interglacial climate. For the past two million years, the Earth has been in an ice age where about 90 percent of the time huge continental glaciers have covered major portions of North America and Eurasia. Roughly every 100,000 years, conditions on the planet dramatically warm, initiating interglacial climates. This is what happened 10,000 years ago. With the interglacial climate came longer growing seasons and greater precipitation on continental landmasses. These two changes brought forth conditions that ushered in a new form of human culture—agriculture.

With the advent of agriculture and improvements in agricultural technologies, global human populations grew steadily for thousands of years. Then with the onset of industrialization and the harnessing of coal energy in the late eighteenth century, we reached a global population of one billion people around 1860. A few decades later, with new medical technologies (particularly germ theory and the development of antibiotics) infant mortality was greatly reduced, and people started living longer. Coupling this with the development of oil as an energy resource that could boost agricultural productivity, global population growth took off and doubled to two billion by 1930. Where it had taken 150,000 years to reach one billion people, a second billion was added in just seventy years. At that time the global per capita consumption of energy was equivalent to 3.32 barrels of oil per year.[3]

By 1979 global populations had more than doubled again, growing past four billion; the global per capita consumption of energy had risen to 11.15 barrels of oil per year.[4] This means that in the span of a half-century our collective transformations of energy, and associated entropy, increased seven fold—a rate that simply isn't sustainable. In fact 1979 marks the highest level of global per capita energy consumption; since that time per capita consumption has fallen as the development of new energy sources has been outpaced by a global population that now stands at about 6.5 billion.[5] Still, the total amount of energy transformed by humans steadily rose and continues to increase today, as does associated biospheric entropy.

Every environmental problem we witness today is the result of entropy within the biosphere. If there is a foundation on which all environmental degradation rests, it is entropy generated by the ever-increasing transformation of energy by humans. The loss of natural forest cover or its replacement with monocrop plantations results in simplification of ecosystems—entropy. The conversion of semiarid woodlands to desert through overexploitation results in ecosystem simplification—entropy. The erosion of topsoil results in diffusion of nutrients—entropy. The eutrophication of aquatic and marine environments from the diffusion of nutrients results in decreased biotic diversity and ecosystem simplification—entropy. The depletion of the world's fisheries results in ecosystem simplification—entropy. The loss of coral reefs and boreal forests due to the warming of oceans and polar climates results in ecosystem simplification—entropy. The loss in global biodiversity results in simplification—entropy.

Global climate change due to the build up of carbon dioxide in the atmosphere from the burning of fossil fuels is a process of diffusion of carbon—entropy. Think of any environmental problem and you will see it is a process where complex systems are being simplified or concentrated materials are being diffused.

Global Climate Change

This last environmental problem mentioned above—global climate change—is perhaps the most serious. Global climate change is the preferred scientific designation for global warming. Yes, the biosphere is heating up, but many residents of the planet are not necessarily experiencing hotter climates. What most people experience are weather patterns that are becoming increasingly more erratic and weather swings that are more dramatic.

There is no longer a debate in the scientific community about global climate change. The evidence from warming oceans and the associated expansion of the water column that is pushing ocean levels higher; the die-off of coral reefs due to the warmer waters; the melting of glacial ice world-wide; boreal and polar regions witnessing intense warming; the increase of dramatic weather events in the form of heat waves, violent storms, floods, and droughts; along with a warming atmosphere, all confirm climate change. The debate about global climate change is no longer centered on whether it is happening or if humans are responsible. We *know* that it is happening and that our actions are an important cause of it. The debate has now turned to what will be the degree and nature of global climate change through the twenty-first century.

A Swiss report noted that in 2003 disasters in the form of droughts, floods, and violent storms cost seventy billion dollars, and that the costs from similar weather events are predicted to more than double during the next ten years.[6] This now seems like a very conservative estimate with the expected costs related to Hurricane Katrina alone topping 200 billion dollars. On October 18, 2005, I woke to hear the startling news on NPR's *Morning Edition* that tropical storm Wilma had overnight exploded into the most powerful hurricane ever recorded in the Atlantic. Seasoned weathermen reported that they had never experienced such an event—a tropical storm with seventy-mile-per-hour winds growing into a hurricane with 175-mile-per-hour winds in less than twenty-four hours. The 2005 hurricane season broke every record in the books: the record number of named tropical storms at twenty-six, the record number of hurricanes at fourteen, and the record number of Category 5 storms at three. Previously the record number of tropical storms and hurricanes were twenty-one and twelve respectively. Dramatic and costly as the 2005 hurricane season was, possibly the most serious threat of global climate change may be the thinning of the Arctic Ocean's sea ice.

The Arctic Ocean has been covered with sea ice continuously for at least two million years. Today it is thinning, from an average thickness in 1976 of 3.1 meters to an average thickness of 1.8 meters by 1997.[7] Not only that, but during this time period the summer coverage of Arctic sea ice has decreased by 23 percent, mostly along the northern coast of Russia, for an average 9 percent reduction in ice cover per decade.[8] This of course creates a positive feedback loop. With sea ice present, a large proportion of summer sunlight currently striking the Arctic region is reflected back out to space.

With an open ocean most of that solar energy will be absorbed, pushing the warming even further.

Historical climatic evidence shows that decreases in water density at the surface of the North Atlantic from greater influxes of freshwater can stall the Gulf Stream. Eight thousand two hundred years ago, due to increased rates of glacial melt-water entering the North Atlantic as the present interglacial was warming up, the Gulf Stream stalled and pushed the global climate from interglacial back into glacial conditions for almost 400 years. The change to glacial conditions was rapid occurring in the span of just a few decades.[9] There are real fears that an open Arctic Ocean will further increase warming of the region, resulting in greater melting of glacial ice in northern Canada, Greenland, and Iceland that could stall the Gulf Stream. If an open Arctic Ocean leads to such an outcome, the consequences for Eurasia, and the entire global climate, would be catastrophic. Europe's moderate climate would be lost and replaced by conditions more akin to Siberia. The Earth would most likely once again enter another 100,000-year-long glaciation. This wouldn't mean that we would see huge continental ice sheets marching southward—that would take thousands of years. But it would mean that growing seasons would decrease worldwide as would precipitation over land. The result would be dramatic reductions in global food production.

The potential threat is so grave that Andrew Marshall of the Pentagon commissioned a report, which was leaked to the press in February 2004, on the national security threats related to global climate change. Marshall, 82, is a Pentagon legend and often referred to as "Yoda" by insiders. He is deeply respected in the Pentagon community, and his association with the report carries a lot of weight. The report's assessment is based on a worst-case scenario—one where nothing is done to curb carbon dioxide emissions and the worst possible results develop. Under such a scenario, the current national security threat from terrorism would look like child's play in comparison. The report states, "Disruption and conflict will be endemic features of life . . . Once again warfare would define human life."[10]

Global conflicts would be fueled by competition for depleted food and water resources, brought on by climate change and associated extreme weather events. The have-nots would be forced into aggressive actions in attempts to secure these necessities of life. Famine would decimate populations in the subtropics and densely populated countries like China and India. The global economy would most likely unravel, and the United States would take on a fortress mentality. The threat of nuclear conflict would be grave.[11] Such a report coming out of the Pentagon sends a powerful message. On a global basis we have exceeded carrying capacity; global climate change is one outcome of a degraded biospheric system. Is this the negative feedback that will correct our overshooting of carrying capacity?

If we do nothing to curb increasing entropy in the biosphere, of which global climate change is just one component, then I think these kinds of outcomes are possible in the coming decades. However, we need to remember the Pentagon report is based on a worst-case scenario. Each year it becomes increasingly clear that climate change is a real threat. Many nations, some U.S. states and cities, and even some corporations are already taking actions to reduce carbon dioxide emissions. It is only a matter of time before the United States will fall in line with all other nations to act. We have the means to dramatically increase energy efficiency and to cut unneeded energy utilization. We only need the collective willpower to do so. This will require some sacrifice of convenience in affluent nations like the United States, but the sacrifice would be minor in comparison to doing nothing.

Just about every homeowner has fire insurance. Although the odds that a fire will occur are very low, the potential risk if fire does occur is very high. Shouldn't we have an insurance policy for the potential risk of global climate change? In this case our only insurance is action.

Personal and Collective Action

Our current notion of progress is based on the idea that to keep progressing we will need to use ever-increasing amounts of energy to stay ahead of resource depletion and waste generation. This is where our paradigm of progress collides with the second law of thermodynamics. From a global perspective, if we continue to use more energy, we will continue to increase entropy within the biosphere unless we can bring in new energy from outside the biosphere to counter that entropy. But where would that new, extra-terrestrial energy come from? Since the level of solar gain isn't seriously increasing, as we transform more and more energy within the biosphere we create greater entropy. There is no way around this.

To replace ecosystem functions compromised by entropy, water purification for example, means consuming even more energy, resulting in further entropy. This becomes a positive feedback loop that just makes things worse. We can't progress under a scenario that produces more entropy in an attempt to fix entropic problems. The second law dictates that our current march toward progress will eventually collapse under its own entropic weight.

Our only solution to counter increasing biospheric entropy is to reduce global energy consumption. Renewable energy resources can definitely help, since they create a lot less entropy than nonrenewable forms of energy, but they have some entropic costs, too. How much entropy results from the mining, refining, and shipping of materials to build solar collectors, electrical wires, and batteries? How much then results from the manufacturing of these things, packing them for transport, the shipping to distribution centers, and finally to stores? Within each of these activities are numerous energy transformations. The wisest approach is to conserve energy as much as possible through the development of our most efficient technologies, and to reduce and eventually cease frivolous, unneeded energy consumption.

Most of the energy consumption that we are responsible for is hidden from view in the extraction of materials, manufacturing processes, and transportation. The energy needed to make a two-gram microchip is equivalent to about a third of a gallon of gasoline.[12] To put that fact into perspective, with the same amount of gasoline and my chainsaw I can cut down and buck up one third of my winter heating supply. I annually burn three cords of wood a year with a cord being a stack that is four feet high by four feet wide and eight feet long. What amount of energy transformation, and associated entropy, lie hidden in the 367 billion microchips produced in 2003 alone?[13] How much entropy results from the need to constantly upgrade computers every few years to deal with ever-changing software? Are such rapid rates in increasing levels of resource extraction and the development of solid waste related to fast-paced changes in computer technology truly necessary?

Increasing biospheric entropy, in all of its forms, including the serious threats of global climate change, is an unsettling reality that can produce anxiety. This recently became clear to me during a visit to a sugarbush owned by John Elder. John is a professor of literature and environmental studies at Middlebury College who enthusiastically manages his forest to sustainably produce maple syrup. We were discussing the implications of climate change and its potential impacts on Vermont's forests, particularly sugar maple, with students from one of his classes. Many projections show Vermont's climate warming enough during this century to drive the sugar maple out of the state. During this impromptu discussion one of the students stated that it was probably too late—that even if we took actions now, the battle to save Vermont's sugar maples was already lost. This assessment was demoralizing to other students who felt powerless by the prospect that global climate change was inevitable. Since there will be a lag response in the meteorological system to any changes that we make, the loss of Vermont sugar maples may be inevitable, but in fact we really can't know exactly how climate change will play out.

When people get anxious they often develop a polarized world-view where things simply become either good or bad, black or white, right or wrong, win or lose. For anyone who adopts this view, loss takes on the sense that "it's all over," even before we really know what any outcome will be. However, in a complex system such simple, polar notions don't capture the true reality of a situation.

In terms of biospheric entropy we are on a continuum. At one end is maximal environmental quality and human well-being; on the other end is serious environmental degradation and human suffering. Right now we are moving toward increasing global environmental degradation. But we will not be able to continue in this direction forever. At some point there will be a turnaround, and we will start moving back toward global environmental quality. The critical point isn't if we are winning or losing, but where will we be on the continuum when that turnaround occurs. The farther we find ourselves from ultimate environmental degradation when things turn around, the better off we, and all the biota of this planet, will be. This becomes the task for anyone interested in environment quality—to slow our current movement down the continuum. In this light every action we take in this regard is a victory. One way to confirm this is to consider what the world would be like today if no environmental battles had ever been fought during the last four decades. In the United States there would be no clean air act, no clean water act, no wetlands protection laws, no safeguards over the production and release of toxic chemicals, and globally no ban on the production of ozone depleting chemicals such as chlorofluorocarbons. I can assure you we'd be much farther down the continuum today, and there would be a lot more human suffering because of it. Even an environmental battle that is lost after twenty years of hard work has been a victory for stalling the movement along the continuum.

There is always time for important action, because any decrease in entropy will also decrease the scale of its impacts. Our task is to seriously decrease our collective consumption of energy. This is particularly true for those of us in the United States, where we consume more energy per capita than any other nation of the world. The average American uses seven times more energy per year than the average global citizen, creating seven times more entropy—a lot of it in gas-guzzling sport utility vehicles. In our personal lives we can make a difference. The more that people dedicate themselves to this task, the greater the difference. We don't need to wait for directions from our leaders; we can do it on our own. In 2003 the Congress voted down a bill to increase fuel efficiency standards for cars made in the United States. Such a bill isn't needed if all people decide on their own to buy fuel-efficient cars. However, legislation can enhance rates of social change.

Large-scale change in complex systems never comes from the top down; it always bubbles up from the bottom. That means that large-scale social, political, and economic change comes from the citizenry, whom elected officials will follow when its collective voice becomes loud enough.

There are so many environmental problems today that to figure out ways to address them in one's personal life can quickly become overwhelming. Many people want to do the right thing but become paralyzed under the complexity and enormity of it all. This is where the second law can be of great help. It is the foundation on which all environmental problems rest. Whenever you are trying to make the right environmental decision, just contemplate which action will generate the least amount of energy transformations. That will almost always be your right choice.

For example, let's say you have the choice of buying tomatoes at a large national supermarket chain or from a local, organic grower. The tomato from the supermarket was most likely grown in California, Florida, or Mexico with lots of fertilizers and pesticides—both of which require a lot of energy consumption to produce. This tomato crop also consumed a lot of diesel fuel to run large farm equipment that tended the crop. The tomatoes were harvested and wrapped in plastic containers—another energy costly activity. Then they were shipped by truck across the country using a gallon of diesel fuel for every few miles traveled. The cost in energy, and entropy, for the production of these tomatoes is huge in comparison to the local, organic grower who doesn't use synthetic fertilizer or pesticides. She also most likely tends her tomato crop without farm machinery once the soil has been turned and seedlings planted. And there are also no energy costs for packaging and shipping. Not only that, the organic tomato will be fresh, juicy, tasty, and far more nutritious than the store-bought tomato that has traveled so far. In this case the correct environmental practice is to buy the locally grown organic tomato. Its price will be higher, but the added price is worth it due to the tomato's better quality, taste, and the low entropic *cost* of its production. Also, if we weren't subsidizing the supermarket tomatoes by passing off the entropic costs of their production to the public and future generations, its price would actually be many times greater than that of the organic tomato. If the price of all products truly reflected the cost in energy consumption related to their production and shipping, energy efficient products would have the lowest prices, and a market system would naturally push for energy efficiency. Such a system would require institutional and governmental policy changes. Those changes would result when the collective voice of citizens pushed for an economic system that fosters environmental and social quality. That collective voice would become possible if our society underwent a paradigm shift.

Due to bifurcation events, large-scale change happens quickly in a complex system. In this regard paradigmatic shifts can race through a culture. Prior to the publication of Rachel Carson's groundbreaking *Silent Spring* in 1962, the words "ecology" and "environment" were probably unknown to almost all Americans. Less than ten years later a large majority of citizens supported environmental action to clean up the country's water and air and to protect endangered species. Due in part to environmental activists organizing Earth Day to increase public awareness of environmental problems, a dramatic paradigm shift occurred in this country in less than a decade. One outcome of this was that Richard Nixon became the most pro-environmental president since Teddy Roosevelt, supporting groundbreaking legislation such as the Clean Water Act, the Clean Air Act, and the Endangered Species Act. He was compelled to do this by a vocal citizenry. The majority of Americans still support improving environmental quality, but the energy for personal and collective action has recently taken a back seat to other concerns.

Right now I sense a growing anxiety in our nation about the future. There are real concerns—terrorism, a politically divided nation and world, job security, access to adequate health care, what appears to be a crumbling educational system, the unsettling aftermath of Hurricane Katrina, the rapidly rising cost of oil—that are all interrelated and bubbling away in our society. They may collectively bring about a dramatic shift in our cultural paradigm—a shift that may portray our current path to progress as being fatally flawed. Such a change could happen quickly. I am hopeful that if it occurs, a new paradigm will develop in our culture in which we will view ourselves as truly being a part of the biosphere, inextricably linked to it, and absolutely dependent on it for our collective well-being. At the very heart of this new world-view would reside a true appreciation for the second law of thermodynamics and an economic system that vigorously fosters energy conservation and efficiency.

Notes

1. John Cavanagh et al., 2002. *Alternatives to Economic Globalization: A Better World is Possible* (San Francisco: Berrett-Koehler Publishers, Inc.), 152.

2. G. Miller, 1971. *Energetics, Kinetics, and Life* (Belmont, Calif.: Wadsworth), 46.

3. Richard Duncan, 2000. "The Peak of World Oil Production and the Road to the Olduvai Gorge." *Pardee Keynote Symposium*. Geological Society of America. See <http://diedoff.com/page224.htmp>, 8.

4. Ibid., 8.

5. Ibid., 8.

6. Natural catastrophes and man-made disasters in 2003. See <http://www.swissre.com/INTERNET/pwswpspr.nsf/fmBookMarkFrameSet?ReadForm&BM=../vwAllbyIDKeyLu/mpdl-5wfeah?OpenDocument>.

7. D. Rothroack et al., 1999. "Thinning of Arctic Sea Ice Cover." *Geophysical Research Letters* vol. 26, no. 23: 3469–72.

8. Daniel Glick, 2004. "The Big Thaw." *National Geographic* vol. 206, no. 3: 21.

9. Kendrick Taylor, 1999. "Rapid Climate Change." *American Scientist* vol. 87, no. 4. See <http://www.americanscientist.org/articles/99arcticles/taylor.html>.

10. M. Townsend and P. Harris, 2004. "Now the Pentagon Tells Bush: Climate Change Will Destroy Us." *Observer* (London) 22 February 2004, p. 1. See <http://www.comondreams.org/headlines04/0222–01.htm>.

11. Ibid., 1.

12. J. Gorman, 2002. "Hidden Costs." *Science News* vol. 162, no. 20: 309.

13. Jason Webb, Semiconductor Industry Association, phone conversation January 12, 2006.

4

THE MYTH OF THE FREE MARKET
The Loss of Diversity and Democracy

"It is time for parents to teach young people that in diversity there is beauty and there is strength." —Maya Angelou[1]

Self-Organization

As they grow, all biological systems increase their complexity, their parts becoming ever more specialized and tightly integrated. As a result the entire system increases its efficient use of material and energy resources, with every by-product becoming a useful resource for some other biological entity. Self-organization is key to the sustainability of all biological systems.

Coevolution and Specialization

In January of 1998 I had the opportunity to co-lead a coral reef ecology trip to the Florida Keys. Coral reefs were an entirely new ecosystem for me and prior to the trip demanded a good amount of research on the various species of fish, corals, and other invertebrates that we would encounter. I also had to become a certified diver. My daughter Kelsey, who was seventeen at the time, got her certification along with me, and we became diving buddies for the trip. I will never forget our first dive within those glorious corridors of coral. The variety of forms and colors of corals and fish were astounding. Even more compelling was how all these species managed to coexist. Each species of fish was specialized to inhabit a particular part of the reef and had evolved a specific means of foraging to avoid direct competition with other species for food. These fish exhibited small, specialized *niches*.

In this case a niche isn't a physical place, such as a small cavity in the coral, but rather an ecological concept that is defined as a species' ecological role. A species' niche involves the totality of its interactions with both the biotic and abiotic components of its ecosystem. It even includes all of a species' behaviors and adaptations. The niche is so all-encompassing that it would be impossible to define it for any organism. Rather than trying to describe a species' niche, ecologists focus on its subsets, like the foraging niche. We are also interested in the dynamics of how niches change—whether they get more expansive or shrink.

Through time, most species' niches grow smaller as organisms become more specialized. This is driven by *coevolution*, a process where species adapt to each other. Outside our kitchen window we have a bird feeder that consistently attracts two species of birds whose niches overlap quite a bit—the black-capped chickadee and the white-breasted nuthatch. Their niches overlap because they live in the same forests and, when they don't have access to a bird feeder, they glean insects off the same trees. They even look similar, with white breasts, gray backs, and black crowns. To reduce direct competition for food, they have specialized in the parts of the tree where they feed, decreasing the size of their foraging niche.

Nuthatches forage on the trunk of the tree while chickadees work the branches. As a result they have developed different beak, foot, and wing morphologies. Nuthatches have a very long back toe and claw. This allows them to "walk" down a tree's trunk. By lifting the back claw they drop a short distance only to stop their descent when they lower the claw to catch the bark. Their much longer beak then gives them the ability to probe bark fissures that chickadees can't access. The chickadee has smaller, rounded wings that allow for short bursts of hovering flight at the ends of branches, where their small beaks glean insects from twigs and leaves. Through these coevolved specializations, chickadees and nuthatches have reduced their competition for food so that they can coexist more successfully.

Competition in nature is quite a bit different than it is in human endeavors. In the natural world species don't seek competition, and more importantly no winners emerge from its struggles. Although an individual or a species may prevail from a competitive interaction, they lose energy during the competition. As a result more energy is lost than if the competitive interaction had never occurred, so they can't be considered winners. Because of energy losses, species move away from competition through time. This is accomplished through the coevolution of specializations that reduce the nature of the competition, such as dividing the foraging areas on a tree.

Just as in competitive interactions, predation also gives rise to coevolved, specialized traits. As predators begin to focus on certain species of prey, the prey develop specialized adaptations to avoid being caught—both are specializations that reduce the size of the species' niches. Summer mornings while having coffee, Marcia and I are captivated by the behaviors of two species—a predator and prey—whose coevolved interactions take place just under the porch roof.

Barn spiders live underneath the shady roof overhang. These are large, hairy, gray spiders that make big orb-shaped webs often more than two feet in diameter. Every morning before sunrise, as the dark of night is drawn away, the barn spiders meticulously remake their webs, damaged from nighttime captures. Yet as soon as the sun comes up they retreat into tight nooks where the roof overhang meets the porch wall. There they remain still, in hiding from mud dauber wasps. The wasps are seeking spiders as food for their young. Any unwary barn spider will find itself being stung, anesthetized, carried away, and eventually encased in a segmented tube of hardened mud. Once the spider is placed in its mud cell, the wasp lays a single egg on it. Upon hatching the wasp larva eats the sleeping spider alive.

Both the spider and wasp have become quite specialized in this predatory interaction. The wasp seeks out only large webs in search of spiders and has developed sophisticated behaviors to trick spiders into exposing themselves. Wasps often feign being caught in a web to lure a barn spider out of hiding. For their part barn spiders have adapted to be mostly active at night, when they focus on moths as prey, and to remain hidden and still during the day. As such the niche of both species has grown smaller due to these specialized, coevolved traits brought on by their predator-prey interaction. Predation and competition are not the only kinds of interactions that generate smaller niches. The coevolution of positive interactions between species also results in specialization, the most striking example being *mutualism*.

Mutualism is a relationship between two species of organisms in which both benefit from interacting and their coexistence is essential for the survival of one or, most commonly, both species. A highly coevolved mutualism, studied by tropical ecologist Daniel Janzen in eastern Mexico, offers a good example. This mutualism involves the bull's-horn acacia tree and a species of acacia ant. Neither species can survive if disassociated from the other.

The acacia tree has developed a number of morphological adaptations to support resident acacia ants. The tree has large spongy thorns—the reason for its name—that can be hollowed out by the ants as nesting sites. To provide food for the ants, the acacia tree has evolved *Beltian bodies*—small, protein-rich extensions on the leaf tips—that can be harvested by the ants without damaging the acacia's leaves. The tree also has open sap wells on its twigs where the ants can access sugary sap as an energy and water supply. Ants removed from an acacia tree starve to death because Beltian bodies and acacia sap are all they will consume. In return the ants vigorously protect the acacia tree they live in from all leaf-consuming animals. Whether it is a small insect or a large herbivore such as a cow, the potent bites of acacia ants drive all unwanted animals away. Not only that, they also cut down any vines that attempt to grow on the acacia tree, and they defoliate branches of surrounding trees that start to invade the acacia's space. In this way the ants offer a super-defense system for the acacia. If the ants are removed from an acacia tree, it will eventually succumb to its herbivores and surrounding competitors.[2]

In this example both ants and acacia have grown so specialized that their very existence is dependent on their mutualistic partner. This may seem like a chancy proposition, for if one goes extinct the other will quickly follow. But coevolution isn't based on future possibilities; it is driven by what is most expedient at the time. Mutualism is such an energy-saving strategy that a large number of species have put all their eggs in this basket.

Coevolution is responsible for two important outcomes in ecosystems beyond reducing the size of species' niches: energy efficiency and species that provide important services to each other. Together these allow for the development of highly integrated, stable communities. As species become more specialized, their efficient use of energy increases. This allows more species to exist in an ecosystem as the finite amount of energy is divided into smaller shares. There is also no waste in an ecosystem; every byproduct released by one species is a critical resource for another. Through decomposition, pollination, dispersal of seeds, and myriad other services species conduct within an ecosystem, they are all inextricably tied to each other's well being. While working to sustain themselves, all species coevolve to support each other. Ecosystems offer a wonderful view of complex, sustainable systems.

Species Richness and Complexity

Through evolutionary time—measured in tens of thousands to millions of years—coevolution decreases the size of niches and as a result allows more species to prosper. During the existence of the biosphere this has resulted in ever-increasing *species richness*—often called *biodiversity*. In *The Diversity of Life*, E. O. Wilson describes five major extinction events that have taken place in the biosphere during the last half a billion years. Each of these events, brought on by dramatic changes in the global environment, extinguished a large percentage of species that existed at that time. Yet in each case, within ten to twenty million years, the biosphere not only recovered its original level of biodiversity of species but exceeded it.[3]

Coevolution is such a potent process that it has produced the conservatively estimated ten to twenty million species that presently inhabit the Earth.[4] The biosphere has consistently increased its diversity and, as a result, its complexity through time.

Increasing complexity is inherent in all natural systems that grow. In chapter two I used the growth of our bodies as an example of an anti-entropic complex system. During our development we witnessed one cell growing into more than thirty trillion cells. Not only that, during that period of growth that one cell differentiated into 254 different kinds of cells—skin, muscle, nerves, blood, and bone cells, to name just a few. Within each cell type there is even further specialization. Some nerve cells only communicate with muscle cells, others only receive messages from sensory cells, and others act as bridges between nerve cells. These differing cells regulate themselves but also work collectively to support each other, just as species in an ecosystem do. This is another example of increasing complexity or self-organization through time—something all multicellular organisms do as they grow and differentiate.

We can also witness the same thing occurring in an ecosystem as it matures successionally. As a forest or coral reef develops following some form of disturbance—such as a hurricane—its species richness increases. In a forest much of this richness is hidden since we have a tendency to focus on large, easily visible species such as trees, wildflowers, birds, and mammals. But in the soil and the canopy of the forest, bacteria, fungi, and invertebrates dramatically increase their numbers through time. All the natural systems we see around us become more complex, efficient, and integrated as they grow.

Factors that Foster Species Richness

Not all ecosystems share the same levels of biodiversity. Tropical rain forest and coral reef ecosystems have much greater levels of species richness than most desert and Arctic tundra ecosystems. To explain why some ecosystems host many species and others relatively few, let us look at the factors that determine the level of species richness in ecosystems.

One important factor that fosters species richness is a stable physical environment, like you would find in an even annual climate rather than one that changes dramatically through the year. Imagine that you were going on two separate yearlong field expeditions—one to Hawaii and one to Baffin Island in the Canadian Arctic. Think about the clothing you would need to pack for each trip. For Hawaii a few pairs of shorts, t-shirts, a sun hat, sunglasses, a rain jacket, a bathing suit, and a pair of sandals would be adequate since the climate is so consistent. But for your trip to Baffin Island you would need extreme cold weather clothing for winter and a wide variety of clothing for the highly variable weather of summer. While a single carry-on bag would do for your trip to Hawaii, you would need trunks filled with clothing for the trip to Baffin Island.

In this analogy the amount of clothing is equivalent to a species' adaptations and the size of its niche. In physically stable environments such as Hawaii's niches can be small since species don't need to have a wide variety of adaptations to handle environmental change. They can also successfully exist with smaller populations since the possibility of a dramatic weather event extinguishing a species is very low. Since niches and population sizes are smaller in stable physical environments, more species can coexist than they can in unstable physical environments that demand large niches and populations.

The amount of physical structure in an ecosystem is another critical factor. While Kelsey and I were diving in the Florida Keys, we at times encountered sections where corals didn't exist. These sites appeared as expansive floors of sand devoid of anything else. Once we entered these sandy areas the diversity of reef organisms such as fish would plummet. The difference in species richness was directly related to the amount of physical structure present in the ecosystem. Where there were corals, the structure of the reef ecosystem was complex. This allowed species of fish to specialize to inhabit different *microhabitats* within the reef. The reef as a whole would be the *habitat* of the fish that lived there, but within it species of fish specialized to inhabit specific parts of the reef—their microhabitat. Many species of fish assorted themselves at different vertical heights within the reef; others specialized to exist with specific species of coral. However, in the sandy flats few microhabitats were available because there was little physical structure, and as a result few species of fish were seen.

A third important factor that fosters species richness in an ecosystem is reduced levels of competition: excessive competition by one species can force out other species in what is termed *competitive exclusion.* There are two major ways that competitive species can be kept in check—through *keystone predators* and *intermediate levels of disturbance.*

A fine example of a keystone predator enhancing species richness comes from research conducted by Robert Paine in tide pools of Washington State. Paine noted that these tide pools normally contained fifteen species of mollusks and barnacles, including the Californian mussel. The major predator of this mussel is the ochre sea star. Paine wanted to observe what would happen to the mollusk/barnacle community if sea stars were excluded from tide pools. Paine handpicked starfish from selected pools while allowing others to harbor starfish as experimental controls. Through this treatment, the species richness of mollusks and barnacles in the pools without starfish dropped from fifteen to eight—close to a 50 percent decrease. Although the number of species dropped, the population of mussels exploded.[5] Without predation by starfish the mussel's population was not kept in check, and it excluded seven other species by out-competing them for space within the tide pool. In this example the starfish is a keystone predator. When removed, the species richness of the tide pool crumbled, just as removing the central keystone from an arch will cause its collapse.

Intermediate levels of disturbance also maintain high levels of species richness. However, too much or too little disturbance can extinguish species. A favorite hike of mine is along the northern section of Franconia Ridge in the White Mountains of New Hampshire. This is a narrow alpine ridge top roughly thirty feet wide and a few miles in length. A three foot wide path snakes along the ridgeline. When I was in college in the late 1960s, on either side of this path one could find healthy communities of alpine plants—dwarf bilberry, diapensia, Bigelow's sedge. Many of these individual plants were centuries old and stood no more than a few inches in height. They existed on this ridge because they could handle the toughest weather in New England—where winter winds greater than 100 miles an hour regularly ice-blasted all other species out of existence. Yet these hardy alpine plants can't tolerate being trampled. Through the last three decades, as hiker traffic dramatically increased on this ridge, more and more people have wandered off-trail to take in the precipitous views. Today on the ridge top, where lush alpine vegetation once existed, one finds only exposed beds of brown gravel. Too much foot traffic created excessive levels of disturbance and a dramatic reduction in species richness in impacted sections of this ridge-top ecosystem.

Too little disturbance can also have the same effect of reducing species richness. To the west of my home is Rocky Ridge. This is a rugged ridgeline with small sections of old-growth forest dominated by eastern hemlock. Today the major disturbances on Rocky Ridge are occasional loggings (in less rugged parts of the ridge) and blowdowns. Hypothetically, if we could put a dome over Rocky Ridge and exclude all forms of disturbance like logging and wind, in time its species richness would dramatically decrease through the competitive exclusion created by the hemlock.

Hemlock is the most competitive tree in this part of New England. It creates very deep shade and a dense, acidic needle layer on the forest floor in which few other species can successfully grow. Without disturbance on this ridge to keep their numbers in check, the hemlocks would soon extinguish most other species. Thus intermediate levels of disturbance are not too great to extinguish species and keep competitive exclusion from occurring.

INTO THE FOREST

At first sight the Pisgah old-growth doesn't appear to have a high level of species richness. I can see only five species of dominant canopy trees: hemlock, white pine, beech, black birch, and red oak. Species of understory shrubs are even fewer, including mountain laurel, witch hazel, and maple-leaf viburnum. Due to the dense shade of the hemlocks, ferns and other herbaceous plants are uncommon. But this view of the forest's diversity is skewed—what one of my colleagues calls "above ground bias." Within the soil and the decaying hulks of downed trees, the diversity of species has become enormous.

The summer of 2003 was one of the wettest witnessed in central New England. During September of that year I visited the old-growth, and what I saw then was completely different than what I am seeing on this hot July afternoon in 2005. That September the understory of this forest was transformed into a fairy-tale land of fungi.

Huge colonies of golden chanterelles covered the ground. Amanitas of various hues sported huge, warty-topped toadstools. Varieties of coral fungi added many textures and colors to the mix. Numerous species of polypores and puffballs festooned the decaying trunks of downed trees. I counted over forty species of fungi growing out of the soil and decaying wood. Two were completely new species that I had never seen before, including a light green, gilled mushroom. Many of the species I saw were saprophytes—such as the numerous species of puffballs—meaning that they get their energy from decomposing wood or other organic material in the soils. There were also many *mycorrhizal* species, such as the amanitas.

Mycorrhizal fungi and trees offer another example of mutualism. The fungus gets its energy from the roots of trees while allowing trees to dramatically increase their rates of nutrient uptake. This is such a critical interaction that many coniferous species of trees can't exist without their mycorrhizae. Part of the mycorrhizae's function is to more quickly decompose organic litter and transfer the resulting nutrients directly to tree roots. In this way mycorrhizae not only help the trees with which they are associated, but the entire forest system by recycling nutrients that might otherwise leach out of the soil and be lost from the old-growth forest. That this forest is able to sustain itself is in large part due to the mycorrhizae and other decomposers in the soil.

In front of me is the rotting trunk of a one-foot diameter beech. It is covered in charcoal mat fungus, making it look like it has been burned, when in reality it is just in advanced stages of decay. I roll part of the trunk back and find two species of salamanders—the robust yellow-spotted salamander and the delicate red-backed salamander—and a whole host of scurrying beetles. All the downed wood in this forest adds greatly to its species richness. Like the reef, the decaying wood creates microhabitats for salamanders, beetles, and many other organisms that I can't even see. The dead beech is a critical resource for these species, which in turn recycle the beech's nutrients to support new generations of beech like the smooth-trunked tree to my right.

This beech has a unique pattern on its bark. I see a series of light tracks, about a quarter of an inch wide, that run all over the bark, intersecting in a grand web. These are the foraging trails of slugs that graze the algae that grow on the beech's bark; in this way the slugs are not unlike periwinkles that graze algae on intertidal rocks. In this forest mycorrhizal fungi support a beech that in turn supports the algae that sustain slugs. When the beech dies it will sustain a whole new community of species. This is just an infinitesimal part of the interlocking web of interactions that make and sustain this old-growth forest. If any species became dominant enough to exclude lots of others, for example if one species of fungus out-competed all the others, the forest would become simplified. In this instance the forest would lose redundancy within its mycorrhizal services, making the forest more fragile. This simplified forest would be less able to resist perturbations that would jeopardize its sustainability if its one species of fungus were extinguished. Certainly it would be far less efficient in its use of energy and would have a greatly reduced capacity to recycle its nutrients. As Maya Angelou states, for this forest its strength is in its diversity.

Corporate Mergers, Competitive Exclusion, and Simplification

So how do increasing diversity in ecosystems and increasing complexity in other natural systems relate to progress? The answer is that systems with more complexity, like ecosystems with higher biotic diversity, have greater resistance to perturbations—they are more stable. As such they can prosper in tough times whereas more simplified systems can start to unravel. In an ecosystem with low species richness, the loss of a single species may cascade through the entire ecosystem, disrupting its function. In an ecosystem with high species richness, the probability that the loss of a single species will result in a similar level of disruption is much lower due to greater redundancy in species' ecological roles.

Every four years, when lemming populations crash in regions of the arctic, major disruptions move through the Arctic ecosystem. Because this ecosystem has low species richness, the loss of the lemming means a huge loss in prey for predators such as the Arctic fox, snowy owl, and gyrfalcon. As a result these predator populations suffer a crash as well. With reduced predation the lemming population quickly rebounds, only to suffer another die-off four years later. Each time the lemming population peaks, Arctic plants they forage are dramatically reduced, allowing soil nutrients to be leached away at higher rates.[6] This biological dynamic increases the instability of the Arctic ecosystem. Ecosystem instability brought on by regular, dramatic cycles is not witnessed in systems with high species richness.

Democracy fosters diversity in a political system. In a healthy democracy that encourages freedom of speech and the press, a broad range of political views are expressed. This generates social stability through openness and accountability, while allowing the society to develop through the evolution of its values. Workers' rights, women's suffrage, civil rights, and environmental legislation are all examples of the evolution of social values during the twentieth century. In contrast, totalitarian states that restrict freedom lack a diversity of views. They represent simple, static systems that can only maintain social stability through brutal military might. Compared to democratic societies, totalitarian states have far more frequent upheavals, coups, and military conflicts, making them more unstable political systems. The foundation of sustained progress lies in stable systems that increase diversity through time to resist perturbations. Is this what we see occurring in our local, national, and global economic systems?

What is seen in the economic arena is just the opposite; a trend where a diverse array of local or regional commercial enterprises have consistently been replaced by ever-larger transnational corporations that increase their size through mergers, acquisitions, and competitive exclusion. A look at the agricultural sector shows this trend quite clearly. Each decade in the United States, the number of farming operations shrinks as more and more of the nation's agricultural production falls into the hands of a decreasing number of agribusinesses that control a constellation of large, industrial-strength farms. Between the end of World War II and 1997, the number of farms in the United States steadily dropped from 5.6 to two million while the average size of farms increased from 200 to almost 500 acres.[7] The year 2002 witnessed a further 1 percent decrease in the number of American farms.[8] Interestingly, congressionally developed farm subsidies that were intended to shore up small family farms are doing just the opposite: In 2003, 60 percent of the subsidies went to large agribusinesses that represented only 10 percent of American farms.[9]

Each of these large industrial farms produces a limited number of crops—low diversity—and consumes great quantities of petroleum (in the form of pesticides, fertilizers, and diesel fuel) to run big farm equipment and to transport its products throughout the world. The Institute on Energy and Man has accurately tracked global per capita energy consumption and oil production through the last century. Global per capita energy consumption peaked in 1979 and has since fallen steadily. The institute predicts that global oil production will peak in 2006 and then will fall more than 60 percent during the next four decades.[10] How will these large corporate farms fare in an environment with dwindling availability of oil? Does it make more sense to support a system in which a larger array of small, diversified farms can be productive on lower levels of petroleum consumption and can market their products locally rather than needing to truck them across the United States? Every example from the natural world suggests that the move to consolidate farms into larger operations increases the threat of future instability. Coupled with global climate change, large operations that are not diversified run an even higher risk of being affected by erratic weather events such as droughts than do smaller, diversified farms.

Now let's see how consolidation affects the quality of our manmade environment. Imagine you are driving down an urban strip at night on the outskirts of any city in the United States. Could you tell what state you are in? Urban strips have become so homogeneous during the last thirty years that it can be hard to know where we are geographically. All strips will have a handful of similar gas stations—Exxon, Mobil, and Texaco. They share the same fast food restaurants—McDonalds, Burger King, Taco Bell, Pizza Hut, and Kentucky Fried Chicken. And many have the same box stores—Wal-Mart and Home Depot are just two examples. Just a few decades ago, regional differences on emerging strips were clearly expressed in the restaurants, clothing, hardware, and other retail stores that were owned and operated by area residents, and they all had a local flare. Those regional differences have been swallowed up by ever-increasing corporate homogeneity, resulting in dramatic losses of commercial diversity. This loss of commercial diversity has broad implications that go far beyond just choices of where to shop.

To see the impact of consolidation of commercial enterprise, a look at Wal-Mart is instructive. Wal-Mart is the single largest business in the world today. Its corporate niche is huge, covering the sale of a just about any imaginable retail product. It is larger than ExxonMobil, General Motors, and General Electric. It is even larger than its top five competitors combined—Target, Sears, JCPenney, Safeway, and Kroger. Out of every consumer dollar spent in the United States during 2002, 7.5 cents went to Wal-Mart.[11] Wal-Mart's hallmark is expressed in its slogan: "everyday low prices." No one can sell products for less than Wal-Mart. This is the reason that Wal-Mart has been so successful. But as the second law of thermodynamics points out, we can never get something for nothing. Low prices come with a number of costs.

Since Wal-Mart captured such an important percentage of the retail market, a large constellation of product suppliers have been caught in a "damned if you do, damned if you don't" situation. To survive they need to contract with this retail giant, but contracting with Wal-Mart often generates loss of control over a supplier's business, losses in profits for the supplier, and sometimes even bankruptcy.

Wal-Mart can dictate what it is willing to pay for a product and often demands that suppliers sell to Wal-Mart at a price lower than what they agreed to the previous year. As a result suppliers have to find ways to manage on lower profits. If they don't, they may be dropped by Wal-Mart. In the case of an umbrella supplier that asked for a 5 percent price increase on their product, a Wal-Mart representative said, "We were expecting a 5 percent decrease. We're off by 10 percent. Go back and sharpen your pencil." The umbrella supplier reworked the numbers and returned with a request for a 2 percent increase. Wal-Mart's reply was, "We'll go with a Chinese manufacturer."[12] The umbrella supplier was dropped.

A common way for suppliers to stay with Wal-Mart is to cut costs through outsourcing—sending manufacturing outside of the United States to places like China or Mexico. Garment workers in China make less than twenty cents an hour, a rate that is impossible to match for American workers.[13] In 2004 Levi Strauss, in its attempts to survive as a business and supply Wal-Mart with an inexpensive line of jeans, closed its last factories in the United States, laying off 2,500 workers. Just two decades earlier it had sixty plants operating in the United States.[14] Cutting employee benefits is another way to cut costs. And suppliers aren't the only ones that cut benefits for their workers. In 2002 only 38 percent of Wal-Mart's employees had health coverage.[15] The largest corporation in the world doesn't offer health plans or retirement benefits to the majority of its workers. This doesn't only affect the employees. One reason health insurance rates continue to increase so dramatically is due to the large percentage of working families in the United States that don't have healthcare coverage.

As I write, three of the largest supermarket chains in the United States—Safeway, Kroger, and Albertsons—just settled a five-month strike with California workers. The deal keeps current workers' wages and health benefits in place, but dramatically reduces both for new hires. The chains want to cut labor costs by one billion dollars to remain competitive with Wal-Mart. When we examine the trend in healthcare benefits given by private industry, we see a drop from 97 percent supplying healthcare benefits in 1980 to 73 percent in 2005.[16] I can imagine that most people would not see this as a sign of progress.

Low prices at Wal-Mart drive quality manufacturing jobs out of the United States, increasing job insecurity in this country; force reductions in employee benefits and wages; and increase pollution and other environmental problems associated with manufacturing goods in less environmentally regulated countries such as Mexico and China. Wal-Mart's low prices have definite costs that are borne by all of us.

Regarding consumer prices, Steve Dobbins, president of Carolina Mills, states, "We want clean air, clean water, good living conditions, the best health care in the world—yet we aren't willing to pay for anything manufactured under those conditions."[17] Decreasing prices of consumer products are used as an indicator of progress. But are losses in job security, job quality, employee benefits, wages, and environmental protection signs of progress? These are costs that as a society we all pay, and they have big impacts on the quality of life for many Americans.

One result of the outsourcing of jobs is that the labor market becomes increasingly unstable. In my parents' generation, many individuals spent their entire working careers with a single organization. There was a higher degree of loyalty between workers and the businesses employing them than we see today. This created job stability that was translated into stable family economies. Today, the average employee's tenure with an organization hovers around 3.6 years.[18] We see that with increasing competitive exclusion in business (resulting in homogenization and simplification of the economic sector) comes increasing instability in jobs and families. Experts tell us that this is just a short-term trend in globalization and that more and better jobs are on their way. Such statements are based on an inherent belief in an ideology of free trade and open markets. Although free trade and open markets are great for corporations, the reality for the majority of individuals and families is that access to quality jobs and employment security have decreased steadily for decades. Why shouldn't we imagine that the new, quality jobs, brought on by globalization, will simply be outsourced, just as the jobs in the information technology sector are?

Another area where we have witnessed increasing consolidation is in the media. This trend is a threat to a healthy democracy because it decreases the scope of views to which citizens are exposed. In his book *The Media Monopoly*, Ben Bagdikian was labeled an alarmist for pointing out that fifty corporations controlled the U.S. news media in 1983. An update of his book in 1992 highlighted that 90 percent of the mass media, including the news, was controlled by less than two dozen corporate giants. He went on to predict that it wouldn't be long before this number decreased to half a dozen, which it did in the year 2000.[19] Not only have we seen consolidation of media giants—AOL/Time Warner, Disney, and Viacom—but major television news outlets are owned by huge corporations. For instance, NBC is owned by General Electric. As more and more of the news we see, read, and hear is produced by fewer and larger for-profit corporations, the depth and diversity of views is reduced greatly to increase profits and reinforce corporate interests. As Lowry Mays, CEO of Clear Channel, which owns 1,200 radio stations nationwide, once stated, "We're not in the business of providing news or information. We're not in the business of providing well-researched music. We're simply in the business of selling our customers' products."[20]

When such a large component of public opinion is formulated through a handful of radio and TV networks, these trends are clearly counterproductive to supporting a vigorous democracy. What then are we to make of the 2004 ruling by the Federal Communications Commission (FCC) to loosen regulations to allow further corporate consolidation of the airwaves? By a three to two vote, the FCC increased the ceiling for ownership of television stations from 35 percent of the national audience to 45 percent and lifted a ban that prevented media companies from owning both newspapers and television or radio outlets.[21] Not only that, but FCC Chairman Michael Powell refused to hold more than one meeting for public comment. Luckily, due to public outcry Congress stepped in and overturned the FCC's ruling. If the charge of the FCC is to oversee the use of the airwaves for public good, why would they rush to push through a ruling to consolidate media clout while at the same time seriously limiting public comment? The answer lies in a trend that has increased corporate power and political influence since prior to the Civil War.

The Rise of Corporate Power

When the United States was founded, as a means to maintain a vibrant democracy the government exerted strong controls over corporate activities. For example, corporations could be founded to engage in public works projects like building a canal, but once the project was completed and the investors had received a decent profit, ownership of the project reverted to the government. Corporations were also excluded from being involved in the political process. But through the mid-1800s, corporate power grew with increasing industrialization and began to influence the political process to its own advantage, eventually resulting in a landmark Supreme Court case in 1886. In *Santa Clara County v. Southern Pacific Railroad*, the court ruled in favor of the railroad. Many people believe that the court decision gave the railroad, and all future corporations, the rights of persons and, as such, coverage by the Bill of Rights. But the notion of "corporations as persons" was not actually in the court's decision. It was added by the court reporter who wrote the introduction—what are known as headnotes—to the decision. Although headnotes have no legal standing, the right of personhood for corporations has been accepted by the Supreme Court since 1886.[22] Included in these rights is freedom of political speech. The Supreme Court later expanded corporate political free speech in the 1970s, supporting corporations' ability to contribute to political campaigns.

More recently, the court's interpretation of the Constitution's Interstate Commerce Clause has also been quite favorable to corporate interests. This clause was originally developed to keep states from obstructing the movement of goods or people across state boundaries. However, corporate-friendly interpretations now allow corporations to challenge most state laws regulating activities within their borders.

In Vermont, where the agricultural economy largely rests in small, family-owned, dairy farms, there was a lot of concern in the early 1990s regarding the use of bovine growth hormone (BGH) to boost milk production in cows. The concern was threefold: worries about synthetic hormones entering the food supply through milk products; fears that, in a BGH market environment, small Vermont farms would become less competitive than big western dairies; and also concern over the well-being of dairy cows who have decreased life expectancies when treated with BGH. These concerns soon were mobilized into a grassroots citizens' effort for a state law that would require all milk products produced with BGH to be labeled so consumers could decide whether or not they wished to buy these products. The bill passed the Vermont legislature in 1994 and was quickly in the courts. The federal court overturned the law in 1996 because it violated a corporation's First Amendment Rights—the right of free speech. In this case it was actually a right "not to speak" so as to withhold information from citizens. That corporations can quash a state's sovereign right to enact legislation that allows its citizens to be better-informed is a scary prospect for democracy, but the Vermont example is just the tip of an iceberg of corporate rule over the sovereign rights of peoples and elected governments.

In 1999 the world got its first glimpse of mass demonstrations against economic globalization as protestors attempted to close down a meeting of the World Trade Organization (WTO) in Seattle. The media generally portrayed the protestors as part of a misguided fringe movement, and globalization as a necessary good. What the media coverage failed to show was that the protesters weren't against global trade; rather, they were protesting current trade agreements that gave corporations power over sovereign governments and peoples (and as such allowed corporations to trump many social and environmental laws developed by democratic institutions, stating that these laws were barriers to free trade).

Under the North American Free Trade Agreement (NAFTA), any corporation can sue a government directly for what it sees as a barrier to free trade. During the spring of 2004 a secret tribunal of the WTO was examining a case brought against the United States by the Canadian corporation Methanex, which produces the gasoline additive MTBE. MTBE has been shown to be a neurotoxin that can be inhaled with gasoline fumes and has a high propensity to contaminate groundwater. To protect public health, California banned MTBE as a gasoline additive. Methanex maintains the California law acts as a barrier to trade and is suing for $970 million.

GATS, the General Agreement on Trade in Services, goes even further. While other trade agreements deal in the free trade of goods, GATS deals in open markets for services, even services regarding public health. Any regulatory laws that can be deemed as blocking access to such service markets as health and safety, discrimination, or pollution standards laws can be grounds for suing a nation. In this way, the sovereignty of governments and peoples to look out for their own physical well-being becomes subservient to corporate interests. To the protestors in Seattle and other cities, it wasn't about global trade but about the usurpation of sovereignty and democracy. The demands of the protestors for fair trade where democratic rights are honored, rather than free trade that usurps them, shows that these concerned citizens are not against trade itself but how it is conducted. (The preceding five paragraphs were condensed from a terrific article by Jeffrey Kaplan published in the November 2003 edition of *Orion*.[23])

Currently, the relationship between the people and the economy has been turned on its head. An economy is supposed to serve its people; however, in the world today, people are to serve the economy. This may be why our leaders often refer to us as consumers rather than citizens.

Not only do individual corporations continue to grow, simplifying the economic system, but political power has been concentrated in transnational corporations to such a degree that democracy itself is being undermined. All the international trade agreements crafted thus far have taken place behind closed doors, with no accountability to the public, by appointed officials. The diversity of opinions, social structures, approaches to governance, and cultural traditions around the world are being compromised as transnational corporations consolidate and increase their hold on power. What is being created is a homogeneous global economy that becomes less accountable to people and less able to resist perturbations. In terms of this last point, Hurricane Katrina exposed the soft underbelly of this system and its inherent fragility.

Within self-organized, complex systems, redundancy of function is a common attribute. In Vermont, if we lose one of our insect pollinators, we have many other species that will service our flowering plants. People with brain injuries who lose their ability to speak often find that speech returns as other parts of the brain take on that function. These are both examples of redundancy in a complex system. What Hurricane Katrina exposed was the fragility of our oil supply system. One storm generated a 40 percent spike in the cost of gasoline, almost overnight! Part of this may have been related to price gouging, but a good part of it was related to the fact that a majority of our petroleum refining capacity lay within Katrina's storm track. This is clear evidence of a simplified system that has little redundancy and, as a result, little stability.

Just as the notion that ever-increasing growth will generate further progress has no scientific grounding, our current global economic system is behaving in a way that is absolutely contradictory to the way all natural complex systems function. Where natural systems grow more diverse, integrated, and efficient, with each specialized part working to support the other parts in a stable system, our global economic system is moving in the opposite direction. It is moving toward simplification and homogeneity through competitive exclusion, wasteful use of resources, and lack of integration, with each corporate entity looking out for its own interests—profits—rather than the well-being of the whole system.

Which model would you bet on for long-term sustainability: the model life on Earth presents, which has successfully sustained itself for at least 3.5 billion years through coevolved specialization, integration, redundancy, and efficiency; or our present corporate model, which has functioned for little more than a century through competitive exclusion and inefficiency?

Notes

1. Maya Angelou, 1993. *Wouldn't Take Nothing for My Journey Now* (New York: Random House), 124.
2. D. H. Janzen, 1967. "Interaction of the Bull's-horn Acacia with an Ant Inhabitant in Eastern Mexico." *The University of Kansas Science Bulletin* vol. 47: 315–58.
3. E. O. Wilson, 1992. *The Diversity of Life* (New York: W. W. Norton and Co.), 191.
4. Ibid., 132.
5. R. T. Paine, 1966. "Food Web Complexity and Species Diversity." *American Naturalist* vol. 100: 65–75.

6. F. Pitelka, 1973. "Cyclic Pattern in Lemming Populations Near Barrow, Alaska." *Alaskan Arctic Tundra* 199–215.

7. U.S. Department of Agriculture. National Agriculture Statistics Service. See <http:://www.usda.gov/nass/pubs/trends/farmnumbers.htm>.

8. Ibid. See <http://usda.mannlib.cornell.edu/reports/general/sb/sb991.pdf>.

9. Environmental Working Group's Farm Subsidy Database. See <http://ewg.org/farm/progdetail.php?fips=20000&progcode=total&page=conc&yr=2003>.

10. Richard Duncan, 2000. "The Peak of World Oil Production and the Road to the Olduvai Gorge." *Pardee Keynote Symposium*. Geological Society of America, 3. See <http://diedoff.com/page224.htm>.

11. "The Wal-Mart You Don't Know." 2003. *Fast Company*. See <http://www.fastcompany.com/magazine/77/walmart.html>.

12. Ibid., 5.

13. Juliet Schor, 2002. "Too Much Stuff: Consumerism with a Conscience." *Boston Globe* 8 December 2002, sec. D, p. 12.

14. "The Wal-Mart You Don't Know," 5–6.

15. See <http://www.indyweek.com/durham/2002–05–08/news.html>.

16. Bureau of Labor Statistics. U.S. Department of Labor. See <http://www.bls.gov/news.release/ebs2.nro.htm>.

17. "The Wal-Mart You Don't Know," 6.

18. Employee Tenure Summary. 2002. U.S. Department of Labor. See <http://www.bls.gov/news.release/tenure.nro.htm>.

19. Media Reform Information Center. See <http://www.corporations.org/media/>.

20. "Giving Our Airwaves to the Media Moguls." 2003. *The Nader Page*. See <http://www.nader.org/interest/053103.html>.

21. See <http://money.cnn.com/2003/06/02/news/companies/fcc_rules/>.

22. Thom Hartmann, 2002. *Unequal Protection: The Rise of Corporate Dominance and the Theft of Human Rights* (Emmaus, Pa.: Rodale), 5.

23. Jeffrey Kaplan, 2003. "Consent of the Governed: The Corporate Usurpation of Democracy and the Valiant Struggle to Win It Back." *Orion* vol. 22, no. 6: 54–61.

5

THE MYTH OF PROGRESS

A Need for Cultural Change

"[I] attribute the social and psychological problems of modern society to the fact that society requires people to live under conditions radically different from those under which the human race evolved." —The Unabomber[1]

"The more materialistic values are at the center of our lives, the more our quality of life is diminished." —Tim Kasser[2]

A Hierarchy of Progress

In 1978 Marcia and I started building our house by cutting, hauling, and barking one hundred white pine trees from our property. These pines were used to build our scribe-fitted log home. The two-story saltbox has a passive solar design, with interior thermal mass from the log walls and a large central brick fireplace to capture and hold the sun's heat. We use a stove to complement solar heating and burn, on average, three cords of wood a year. Our kitchen stove, water heater, and backup space heater run on propane. We have thirty acres of woodland that take up more carbon dioxide in photosynthesis than Marcia and I release from the burning of wood and propane annually. In this regard our environmental footprint is pretty good, but we are far from minimalists as environmental practitioners.

We have two vehicles—a four-wheel-drive truck and a Toyota Echo. We have two computers and printers—an iMAC and the Macintosh we bought in 1989. By all standards the older Macintosh is obsolete, but I wrote my first book on this computer, and it still works. I can't bring myself to get rid of it. So I use it whenever I write letters or small documents that I need to print. We have a television that gets no reception, so it is used with a VCR. We have a CD player and radio for music and news. We have a washing machine, but not a drier since we hang clothes to dry. In the kitchen we have a refrigerator, a toaster, a blender, an electric crock-pot, and an electric coffee grinder. I also have a chain saw, a radial-arm saw, a skill saw, a jigsaw, a router, and an electric drill. Marcia, who enjoys yard work, has a lawn mower. Other than lights, the only other electronics that we have are a ceiling fan to circulate air and our water pump. We don't have a cell phone or answering machine for our phone.

We have been pretty conscious about our purchases, particularly of items that consume energy, yet compared to a vast majority of people on this planet we are extremely affluent and our material possessions allow us to live a life of comfort and ease. Even my grandparents, if they were alive today, would be in awe of some of these items—two cars, two computers, a VCR, a CD player! These possessions are clear indicators of what I call *material progress*.

There is no doubt that as a culture we have witnessed material progress that has increased convenience, extended life expectancy, and dramatically expanded communication and information sharing. Most of the indicators used to support our reigning paradigm of progress relate to materialism. Yet,

I believe that materialism is much too limited an indicator with which to mark progress. Much more important trends of progress relate to the physical well-being of citizens, their emotional well-being, and community well-being—what Robert Putnam in *Bowling Alone* labels social capital. Together all form a hierarchy of progress.

Community well-being occurs when citizens make personal sacrifices to help people, and even non-human organisms, that lie outside their circle of family and friends. They are helping directly those they wouldn't otherwise have contact with in their daily activities. A society in which most of its citizens engage in this kind of benevolent activity has truly progressed. Individuals who engage in community outreach have expanded their emotional sensitivities to encompass a much larger sphere than just those they know.

I recently heard a short commentary on Vermont Public Radio by a University of Vermont student. Rather than heading south to a sunbathed beach for spring break, she spent her time in a city she had never visited before to help the homeless. This young woman truly displays an expansive compassion for less privileged people that radiated far beyond her own community to people she helped and most likely will never see again.

In contrast, the emotional attention of some people is focused solely within themselves. In this way their bodies become a cell for their emotional life. For these people everything is judged by "how it affects *me*." These judgments are usually negative, causing individuals to feel that injustices have been committed against them and that they are due compensation. For people like this, other human beings don't really exist—they are simply agents of wrongdoing. Expanding the emotional sphere outward, a person may embrace her immediate family or a tight circle of friends. For these individuals, only family members or close friends are important—people outside this sphere are not of concern. A further expansion may incorporate extended family, friends, and even acquaintances. Then there are those like the UVM student whose attention goes well beyond their family and friends, to people they don't even know. In these individuals compassion and empathy are highly developed, as is their connection to others. Interestingly, similar connections happen in forests between mycorrhizal fungi and trees.

Mycorrhizae are fungi that live in the soil and get their energy from the roots of living trees. These fungi are not parasites, though, because they help trees to dramatically increase their absorption of water and nutrients. Both trees and mycorrhizal fungi benefit from their interactions; for many species of coniferous trees, this interaction is necessary for survival. In 1997 a study was conducted in a Douglas fir and paper birch forest in British Columbia to quantify just how much of the trees' energy was consumed by their mycorrhizal fungi. To do this, carbon dioxide labeled with carbon 14—a rare and radioactive form of carbon—was introduced to the Douglas fir in order to find out how much of the carbon 14 ended up in the mycorrhizae as carbohydrates produced by the trees. An unexpected and rather profound finding came out of the study.

Paper birch trees that were struggling in the forest received as much as 6 percent of their energy from carbohydrates produced by surrounding Douglas fir. The extra energy was fed to the struggling birch by the mycorrhizae.[3] In this way the mycorrhizae acted like tree shepherds helping along struggling members of their flock. The action of the mycorrhizae benefited the entire forest ecosystem by maintaining or increasing photosynthetic rates, nutrient retention, and species richness. Just like the mycorrhizae, our task as individuals is to progress in a manner in which our attention, compassion, and empathy grow ever-outward to benefit our communities and society as a whole.

However, to accomplish this kind of progress individuals first have to reach a state of emotional well-being. They have to feel good about themselves and engaged in their lives. Their experience needs to be *fulfilling*.

I need to make a distinction here between fulfillment and pleasure. Pleasure is an ephemeral experience; once the object that gives it is removed, pleasure fades. Eating ice cream is quite pleasurable, but once the ice cream is gone, so is the experience. Fulfillment, on the other hand, has staying power. Love is a fulfilling experience. Even if our loved ones are not present, their impact stays with us. Pleasure can never lead to emotional well-being because we can't carry it with us, but fulfillment can. In order to foster community outreach, a society has to be developed to the point where its citizens are emotionally healthy and fulfilled by their experience of life.

To be emotionally healthy, most people need to be physically healthy, too. Physical well-being results from a healthy lifestyle—nutritious diet, exercise, and good sleeping habits—and is marked by vitality. Physical well-being is not derived from medical intervention, such as the use of pharmaceuticals. Once medical intervention is needed, physical well-being has already been compromised to some degree.

On the bottom rung of this hierarchy we find material progress. Material progress can foster physical and emotional well-being. But it can also erode them if a society becomes too focused on materialism and affluence. Tim Kasser's research, as well as the research of other psychologists who have looked at the relationship between materialism and well-being, consistently shows that once people are financially above the poverty level, aspirations for increased affluence are associated with decreases in happiness. The more affluent people become, the greater the rates of anxiety, depression, and social isolation. Studies also show that with increased affluence people use more drugs and alcohol and have decreased levels of vitality.[4]

As previously mentioned, the indicators of progress that we most commonly use relate to material progress—GDP, per capita income, and increasing life expectancy—and they are all measures of quantity rather than quality. Let's look at the last indicator: life expectancy. People in the United States are living longer due to medical technology—a form of material progress—but are they living physically healthier lives? I pointed out in the introduction that one out of three Americans is overweight and obesity has seriously increased by more than 400 percent in the last two decades.[5] Asthma rates continue to increase, particularly in children.[6] An estimated 6.5 percent of Americans are diabetic, many of these due to obesity.[7] Heart disease continues to be the number one killer. These are all diseases that have a strong environmental basis. They are indicators of a society in which the physical well-being of citizens is deteriorating, not progressing, and contribute to the annual rise in healthcare costs. A similar trend can be seen in emotional well-being: Skyrocketing rates of environmentally induced unipolar depression and suicide hold as the eighth leading cause of premature death. Yes, we are living longer due to medical interventions, but do these trends suggest we are living better?

When people gain too much material affluence, their focus is often turned to their possessions and the maintenance of an affluent lifestyle. It's true that possessions can bring ephemeral pleasure, but as Kasser's work points out, rarely does consumption lead to fulfillment. When an individual's focus is primarily directed at material wealth, his attention is drawn inward, shrinking his emotional sphere. Even though he may have the financial means to promote community outreach, it isn't necessarily on his screen. In this way possessions can develop ownership over people as they become enslaved to a material-rich lifestyle—an enslavement that can erode both physical and emotional well-being. I contend that our society has focused far too long on materialism as a means to progress, halting real societal progress. Today there is a loud call to rework our educational system to prepare a more sophisticated work force that will be able to handle the new, highly skilled jobs that globalization will produce. I hear fewer voices out there calling for changes in education to promote social capital or engaged citizenship; as a matter of fact, voting rates in the United States have dropped below 50 percent. How is it that a country that is the hallmark of democracy has witnessed the erosion of civic responsibility? I believe that the pursuit of material progress has usurped the sociopolitical ideals on which the United States was founded.

Ancient Cultural Values

In the previous chapters I have applied a scientific perspective to point out why our current path to progress is not sustainable. Here I am stating that it also isn't possible due to societal values that focus primarily on material progress and give rise to ever-increasing consumption. I have pressed the point that real progress will be attained only if we develop a socioeconomic model that fosters diversity and energy conservation, and achieves a dynamic equilibrium in which the amount of materials and energy consumed annually remains the same and can be supported by the biosphere. Such a system becomes possible if we can slow and then reverse global population growth, and develop cultural values that turn us away from ever-increasing consumption and toward progress in physical, emotional, and community well-being.

Economist Herman Daly has already developed a model for a socioeconomic system that functions in dynamic equilibrium. David Korten, in his book *The Post-Corporate World: Life After Capitalism*, goes further in modeling "living economies" that function under the same principles as life's complex systems. Korten's basic attributes of living economies grow out of the concept of self-organization in biological systems—that systems increase complexity, diversity, integration, and stability through time. Korten advocates replacing huge multinational corporations with smaller local and regional businesses that are specifically adapted to the region they serve. If these businesses are publicly owned, the ownership is by citizens of the region. In this way businesses and their shareholders will work for the good of their community and regional environment rather than solely attempting to maximize profit. Businesses should share information and work to support each other rather than engaging in competitive exclusion. This cooperation would result in more specialized and integrated commercial enterprises. Businesses should strive to be frugal and very efficient in their use of material and energy resources. This would not only decrease consumption and waste but allow more resources to be available for other businesses.[8]

These are just a sampling of the ideas that Korten develops in his book. Along with other economists, he shows that sound, sustainable, economic models already exist. But in order to engage in such a socioeconomic system, we will need to embrace a new set of cultural values—what I call *ancient values*. These are not the "traditional values" that some politicians claim will elevate society—values that characterized late nineteenth-century America. Although traditional values stress the importance of family, they also support a rugged individualism that promotes individual entitlement regardless of its social or environmental consequences. A large part of the problems we face today has been spawned by individual entitlement and its self-absorbed focus. In order to really progress we need to look to far older values—ones that existed long before the development of agriculture. To explore these ancient cultural values I recount the following experience.

It's 1993 and I am about four miles north of where we are camped in the Pinacate region of Mexico. It is what we call "solo day"—a chance for students on this Antioch desert ecology field study trip to explore and connect to this unique landscape in their own way. I'm using the day to explore a new area of the Pinacate—the most glorious hot desert landscape that I have ever encountered.

It's been a wet winter and spring, so the desert is lush. Fields of brilliant apricot-colored desert mallows cover black cinder substrates. Older, reddish lava flows are carpeted by the yellow bloom of brittlebush. Most appealing of all is the ocotillo, with its emerald green wands topped by flaming crimson flowers. In most deserts with light substrates, these colors would look washed out during the day, but in contrast to the black cinders and lava flows of this rugged landscape, any color is brilliant. It is the combination of this geologically young, volcanic landscape and its exquisite mix of vegetation that places the Pinacate at the heart of our desert experience.

I crest a ridge formed by an old lava flow, descend into a desert basin dominated by creosote bush, and cross a large arroyo—a dry streambed. As I start to climb out of the drainage I see a pile of lava rocks about 200 feet to my left, up on the lip of the arroyo. I alter my course to check out the cairn. As I approach it, I stumble upon a significant find—something I have previously only read about—an ancient footpath.

The footpaths of the Pinacate link lava-lined water holes called tanks and eventually lead to the Sea of Cortez for the gathering of salt. The path is a distinct trough in the desert floor. Large and small rocks glistening with desert varnish line its sides. Desert varnish is a coating of manganese and iron oxides that ever so slowly coats desert rocks that remain set in place. I pick up one of these rocks; its dark chocolate–colored varnish is as smooth to the touch as enameled porcelain. Such a layer of desert varnish takes millennia to form if the rocks remain fixed in their positions during that time. The varnish confirms that this footpath is thousands of years old. I try to imagine how many generations and how many feet traversing this path pushed the rocks to their present places of rest?

The last native people to walk this path were the O'odham, also known as the Papago. Before them, it may have been the Hohokam. Before the Hohokam, unnamed hunters and gathers lived here for millennia. Varnish-covered Clovis spear points dating back to 12,000 years ago have been found embedded in these footpaths. Based on microscopic inspection of the desert varnish that covers rocks associated with the Pinacate paths, some researchers have pushed the paths' origins back to 35,000 years ago. This assertion has sparked a lively debate, but even if these footpaths are only 12,000 years old, it still makes them the oldest landscape antiquities in North America.

Instinctively, I step onto the footpath and start walking in my thick-soled boots. I see up ahead that the path is going to enter one of the Pinacate's youngest lava flows. The realization stops me in my tracks, because I remember reading that the first Spaniards to encounter the O'odham in the seventeenth century mentioned that they crossed this landscape barefoot. The Vibram soles of my boots are chipped and scraped by just a few days of exploration of the Pinacate's lava flows; one lug has been cut right off. What kind of feet did the O'odham people have? And then, in that moment opened by my question, a second, more profound one arises in my mind: What was life really like for the ancient hunter-gatherers who used to walk these paths?

I'm sure life was physically tough and very hard times were common. Summer temperatures regularly climb to more than 120 degrees; on the black cinder flats ground temperatures can burn exposed skin. During some years this desert region receives less than an inch of rainfall. At such times food and water are scarce, demanding deprivation and long desert treks. The Spanish explorers of the seventeenth century couldn't comprehend why native people chose to live here. From the European perspective, this region of the Sonoran was not only a wasteland but also the very vomit of the Earth—an entirely unwholesome and unclean place. Yet I have a strong sense that even though life was physically difficult and life expectancies were short, the *experience of life* for the people that lived here thousands of years ago was extraordinarily rich. I base this on the following suppositions.

Hunter-gatherer desert culture was based in nomadic clans of a few dozen people. Within the clan group each person had a specific role, and the entire clan group relied heavily on each other and shared all that it had. Like all hunter-gatherer groups, if someone was successful in a hunt, the meat was shared with those who didn't have success. If any individual accumulated too many possessions, a giving-away ceremony took place so that no one individual had too much. In this way, these ancient people practiced reciprocal altruism as a means to survive in this harsh environment. There was no room for personal greed. All individuals had a direct voice in how the affairs of the clan would develop—whether they should move to the next tank, celebrate a particular occasion, or conduct a sacred ritual. For these people the idea of needing to *create* community would have been absurd. They *were* community—on the deepest of levels. Through stories and rituals, in joy and sorrow, they shared the very core of their lives. I believe that this very strong sense of community, where each member was truly an integral part, greatly enriched their experience of life.

Not only did each individual have a critical place within the clan, each individual also clearly knew his or her place within the world. Through rich traditions, in the form of stories, rituals, and sacred practices (all of which had been passed from generation to generation for hundreds, possibly thousands of years), these people were seamlessly woven into their landscape. As hunter-gatherers they saw themselves as a part of the land, not apart from it, sharing it with all the other plants and creatures on whom they depended for survival. Their world made sense—it was truly their home. Even though the desert is harsh, it holds a beauty and mystery that I have found in no other landscape. As a once-a-year visitor I can vividly sense the vitality in this place. It has a deep impact on me, but I can't begin to imagine the depth of the ancients' experience of, and connection to, this land. I am confident that their experience of life was also greatly enriched due to their intimate connection to this place.

Finally, like all hunter-gatherers, they had plenty of time to socialize, tell stories, make crafts, and reflect on their existence. Reflective practice is essential to convert knowledge into understanding and, eventually, wisdom.

Knowledge and understanding are often used interchangeably, but I see them as distinctly different. Knowledge is the acquisition of factual information. It is strictly a mental phenomenon. That our bodies comprise more than thirty trillion cells is a piece of my knowledge. Understanding, on the other hand, is being able to comprehend the meaning or implications of knowledge. Just how many is thirty trillion? In addition to thinking, understanding is characterized by both an emotional and physical response. Where knowledge is black and white, right or wrong—the sort of stuff that is tested for in objective exams—understanding is the many-layered lotus blossom. There is always room for deeper understanding. It runs from the sense of AH HA! depicted in cartoons as a light bulb going off over someone's head, to epiphany, to deep revelatory experience. Where knowledge is static, understanding is dynamic, multifaceted, and always carries with it some level of fulfillment. Understanding is an experience that inflates us.

On the other hand, if we carry too much unprocessed knowledge, it can deaden us. I used to teach a Concepts of Biology course at Antioch. It was a class for students who never had a college-level biology course. The two most common reasons that these students didn't take biology as undergraduates were that they either mistakenly got the impression in high school that they just weren't good at science, or their experience with high-school biology was utterly boring. For me, it's hard to imagine biology as boring. When we start to have a glimmer of understanding regarding the complexity of biological systems and how beautifully they function, it becomes completely engrossing. How could anyone be bored by biology? For the Antioch students the answer to that question lies in high-school courses that, based on a linear mode of instruction, were geared solely toward the acquisition of knowledge through the memorization of endless facts and terminology. Without any opportunity to reflect on that knowledge and translate it into understanding, their experience was deadening.

Reflective practice is not solely based on contemplation; it is also fostered through the arts. Painting, sculpting, composing and playing music are all means of reflective practice that don't involve verbal articulation. Artistic works help process knowledge and directly impact the emotional and physical centers of both the practitioner and the audience. As such, the arts also work for the promotion of understanding. The Pinacate's hunter-gatherers had ample time for reflective practice through their arts, stories, and time for contemplation, which all helped to forge a rich experience of life.

To have ample time for reflection to generate understanding, to be an intimate member of a rich communal life, to know your place in the world through vast traditions, to be intrinsically connected to the land: All these things work to create a rich experience of life—one I'm convinced these ancient people had.

Ten Thousand Years of Cultural Transformation

These sorts of important connections and time for reflective practice are cultural attributes desperately needed today. Our species, modern *Homo sapien*, has existed on the earth for at least 150,000 years. For almost 95 percent of that time all humans shared a mode of life in the form of hunter-gatherer culture. They also shared connection to community, connection to place, and time for reflective practice as the foundation on which their culture was grounded. Why in today's society have these cultural attributes atrophied to such a degree?

Ten thousand years ago, as global climates warmed after the last glaciation and growing seasons lengthened, a new form of human culture evolved—agriculture. Through time the village and extended family replaced the nomadic clan. People continued to have the strong communities, rich traditions, close connection to the land, and ample time for reflection that grounded them in their world. But two important changes emerged with agriculture. The first was that the sense of being a part of the land was replaced by being *apart* from it. The idea of having dominion over the Earth represented in Genesis is a direct outgrowth of agriculture. Secondly, as villages grew in size, political hierarchies developed. This meant that the decision-making process was not equally shared by all. For the first time, many individuals no longer had the ability to be involved in decision making that directly affected their lives and culture.

For thousands of years, agricultural innovation allowed villages to grow and become cities with complex economies and transportation systems, but the development of urban settings (where the majority of the people were disconnected from some form of meaningful relationship with the land) didn't begin until 200 years ago, when industrial culture was ushered in on fossil fuel–driven steam engines. With industrial culture, extended families were shed for more-mobile nuclear ones as the ability to travel via ship, train, auto, and plane became easier. Societal changes accelerated, and coupled with greater mobility, connections to traditions that grounded people to their place were lost—and with them was lost the ability to help people make sense of their world. Even though labor-saving technology made life physically easier, increasing the complexity of lifestyle actually left less time for reflective practice.

And today we find ourselves crossing the threshold into our fourth major cultural transformation. With the onset of global, postindustrial culture, we see dramatic shifts in populations due to political and economic upheavals, plus ever-changing job markets. Some estimates suggest that two billion people, or one out of three humans, have been displaced from their homelands in the past few decades by war and economic systems that have left them behind.[9] Gary Nabhan points out that the words "peace" and "place" have similar roots. Thus true peace and security are linked to being connected to one's place.[10] For people ripped from their homelands, both peace and quality of life have been seriously eroded.

In the United States, where people are not displaced by conflict, the job market has become increasingly prone to perturbations. Partially due to job market instability, by 1996 the average U.S. citizen had moved every 4.7 years.[11] How is it possible to build a connection to community or place when moving so frequently? To make ends meet, the vast majority of American families now have two or more wage earners, and many individuals work multiple jobs. In the mid-1990s America passed Japan to become the nation whose citizens work the longest hours of any country in the world.[12] Because of the impacts of working longer hours, families spend far less time together than they did just a couple of decades ago. Like the extended family a century ago, the nuclear family now finds itself under increasing pressures that threaten its integrity.

Although egalitarian decision-making was eventually lost with agriculture, elected officials in industrial democracies did bear the brunt of the decision-making process that impacted citizens' lives. Today many critical decisions regarding our collective global future are being made behind closed doors by trade representatives—appointed officials—often with the blessings of amorphous, transnational corporations. Never in the history of democratic societies has the populace been more removed from the decision-making process than it is today. The combination of these trends has not only isolated a large part of the populace, but also disenfranchised the vast majority of people in decisions that directly impact their lives, their culture, and the lives of future generations.

With voice mail, e-mail, call waiting, cell phones, and faxes, we are finding more time for "productive" ventures but less time for real involvement with people. For many, DVDs, computer games, sophisticated software, and the Internet are replacing the real world with a virtual one. Yes, we are gaining the sense that we are truly a global community. But is that sense being translated into greater community outreach? As the cascade of information that we are all exposed to grows exponentially, where do we find the needed time to reflect on it and extract understanding of the world around us? Where are our children being exposed to reflective practice when art and music programs are being cut in schools throughout the country and more and more time is spent in cyberspace? As T. S. Eliot writes in *Choruses from the Rock*, "Where is the wisdom we have lost in knowledge? Where is the knowledge we have lost in information?"[13]

In terms of traditions, what most typifies American culture

today? The Super Bowl is a cultural event shared by more people in the United States than any other. When we reflect that a big part of its draw is a showcase for new advertising, we need to ask: Has consumerism become the icon of our culture? Watching the network news on November 25, 2005, suggests that consumerism has become the hallmark of our society. The lead story that night wasn't about the war in Iraq, the federal deficit, or any number of pressing national issues, but rather about Black Friday—the vaulted shopping day that follows Thanksgiving. Just as Good Friday is one of the high holy days of the Christian faith, it appears that Black Friday ushers in the high holy days of shopping. As stated previously, consumption has become such an important tradition that our president no longer addresses us as citizens, but now as consumers. When our country is in crisis, rather than being asked as citizens to sacrifice, we are asked as consumers to shop! We are told that free trade and open markets will benefit the consumer with lower prices, so we will be able to consume even more. A little more than a century ago frugality was an inherent American ideal; today consumption appears to be the focus of our culture.

The Means to True Progress

Of course, the last few paragraphs intentionally cast a decidedly one-sided description of the evolution of our present cultural state of affairs. In reality there are many wonderful attributes spawned by our cultural transformation over the last few thousand years. These include, to name a few, the rise of democratic institutions, advances in the rights of women, the expansion of civil rights, the advancement of scientific understanding, and further development in the arts. Yet, in a singular way, we have become the flip side of the coin from the Pinacate's hunter-gatherers. Whereas their life was physically challenging but experientially rich, everyday experience has become physically comfortable and experientially poor for many Americans today.

Just as our hands and recessed eye sockets are the direct result of our arboreal past, our need for real community, traditions that help us find our way, connection to our place, and ample time for reflective practice is a direct result of our cultural legacy. Since these things are essential to being human, and intrinsically necessary if we are to have a rich, fulfilled experience of life, they are essential if we are to have real progress.

As we have been drawn away from connection to community, place, and reflective practice, a void has developed—what I call a "hollowness of experience." That void is presently being filled by a need to consume. Yet ever-increasing consumption doesn't make us happier or more fulfilled; it does just the opposite.[14] As we have become isolated from community and place, reciprocal altruism and stewardship have been replaced by self-absorption. When we are connected to community and place we care about them, and our actions reflect that caring as we work for their well-being. Without those connections we lose awareness of how our actions impact others or the environment, and without reflective practice we also lose any sense of responsibility for our actions. As such, greed becomes possible and when linked to the need to consume, the combination allows for dramatically selfish behavior. How else can we explain the callousness displayed by CEOs and CFOs of bankrupt corporations such as Enron and World-Com? The isolation of people from community, place, and reflective practice has become a crisis of culture.

To be able to engage in an economic system not based on continued growth, we need to find ways to sustain ourselves that are not based on materialism. Our attention needs to be turned toward fostering community, strong connections to place, traditions that link community to place, and reflective practice to generate understanding and eventually wisdom. These are the only means to bring forth true, sustainable progress for humanity.

Notes

1. Robert Wright, 1995. "The Evolution of Despair." *Time* vol. 146, no. 9: 50. See <http://www.time.com/time/m...e/archive/1995/950828/950828.cover.html>.

2. Tim Kasser, 2002. *The High Price of Materialism.* (Cambridge, Mass.: MIT Press), 14.

3. S. Simard et al., 1997. "Net Transfer of Carbon Between Ectomycorrhizal Tree Species in the Field." *Nature* vol. 388: 579–82.

4. Tim Kasser, 2002. *The High Price of Materialism.* (Cambridge, Mass.: MIT Press), 12.

5. A. Mokdad et al., 2001. "The Continuing Epidemics of Obesity and Diabetes in the United States." *Journal of the American Medical Association* vol. 286, no. 10: 1197.

6. D. Mannino et al., 2002. Surveillance for Asthma—United States, 1980–1996. Center for Disease Control. See <http://www.cdc.gov/mmwr/preview/mmwrhtml/ss5101a1.htm>.

7. National Diabetes Information Clearinghouse. See <http://diabetes.niddik.nih.gov/dm/pubs/statistics/index.htm#9>.

8. David Korten, 1999. *The Post-Corporate World: Life After Capitalism* (West Hartford, Conn.: Kumarian Press, Inc., and San Francisco: Berrett-Koehler Publishers), 121–33.

9. Gary Nabhan, 2004. "Listening to the Other." *Orion* vol. 23, no. 3 (May/June): 20.

10. Ibid., 22.

11. "Seasonality of Moves and the Duration and Tenure of Res-

idence." 1996. U.S. Census Bureau, 4. See <http://www.census.gov/population/www/documentation/twps0069/twps0069.html>.

12. Richard McNeil, 2004. Radio interview on New Hampshire Public Radio's *Exchange Program*. 8 October 2004.

13. T. S. Eliot, 1952. *The Complete Poems and Plays 1909–1950* (New York: Harcourt, Brace and Company), 96.

14. Tim Kasser, 2002. *The High Price of Materialism* (Cambridge, Mass.: MIT Press), 21.

EPILOGUE

From Consumption to Connection

October 1969 found me as a depressed industrial engineering student at Johns Hopkins University. I felt out of place with my program of study and Baltimore as well. Sensing my distress, a fellow dorm mate gave me a book that he said would be a good antidote to the engineering texts I was reading. The book was *Black Elk Speaks* by John G. Neihardt. I can vividly remember the powerful impact this book had on me. Within the first three pages of chapter one, The Offering of the Pipe, I sensed a whole new world-view— one that represented a deep spiritual connection to the Earth. To this day I think the first few pages of this book are the most inspiring prose that I have ever read. What Black Elk presents is wisdom totally different from anything I had been exposed to previously in Western culture. He talks about the holiness of deep connections to the Earth, all the life it supports, and the importance of a people being whole. He doesn't speak of affluence or material wealth, only the importance of connection.

As a young boy Black Elk received a vision meant to help his people regain their wholeness in the face of devastating impacts from white encroachment into Sioux territory. Toward the end of his life, Black Elk felt that he had failed his people by not being able to fulfill his vision. Yet his words carried on, and long after his death they woke something in me, just as they have done in many others. For the first time I started to question some of the values our culture holds dear. Little did I know it at that time, but Black Elk's wisdom would chart a new path for my life and eventually lead to the writing of this book. If *The Myth of Progress* has any success in opening the minds of its readers, it is in part due to Black Elk, who understood the importance of connection.

My guess is that Black Elk also understood, as he experienced the cycle of life, death, and renewal on the prairie, that continual growth simply is not possible. Just as this was true for the reindeer population on St. Matthews Island, it is true for our socioeconomic system.

As the global economy grows and consumes ever-increasing amounts of energy and resources, entropy throughout the globe will accelerate. The feedback from this mounting entropy will eventually destabilize and curtail economic growth. With an economic system that continuously moves away from cooperative integration of efficient, specialized enterprises to huge transnationals that thrive on competitive exclusion, we find a system that grows increasingly wasteful, lacks critical redundancy, and as a result moves toward greater instability. How many empires and civilizations have collapsed because they grew past the means to sustain themselves? It's not a matter of whether this current economic system will fail; it is simply a matter of when it will fail.

I know there will be critics of this book—people entrenched in the neoclassical economic paradigm. They may say that the final sentence in the previous paragraph is simply more "doom and gloom"—a label meant to discredit ideas rather than counter them with rational arguments. In response, I say that it is actually an objective assessment based on scientific law. I encourage critics to explain how we can maintain continued economic growth in light of the laws of limits to growth, self-organization, and the second law of thermodynamics. In fact I'd be relieved to find out that I have erred and that progress through continued growth is actually possible. But sound science strongly suggests otherwise.

Although I am not optimistic about the upcoming decades, I am not pessimistic either. I have hope that we will be able to transition into a new socioeconomic system based on sustainable principles. We have the resources, appropriate technologies, and understanding to make such a transition; we only lack the collective will power to do so. Luckily due to bifurcation in complex systems, a cultural paradigm can shift quickly. My hope is that as all the feedback within our current system mounts—job insecurity, loss of benefits, lack of access to adequate health care, soaring energy costs, weather disasters, environmental degradation, even threats of terrorism—people will realize collectively that our current path clearly is not leading toward progress. It is also my hope that people will come to realize, as Black Elk so eloquently points out, that it is through connection, not consumption, that will allow us to thrive.

I am encouraged by people dedicated to decreasing their ecological footprint, people who support local enterprises and create sustainable networks. These individuals are developing the prototypes of a socioeconomic system that will promote sustained prosperity. In the Connecticut River watershed where I live, encouraging endeavors abound. Farmers' markets have spawned numerous community-supported agricultural (CSA) initiatives, where people buy shares in a farm and get all their produce as members. Some of these CSAs have linked up with other specialized farms to offer dairy products, eggs, locally raised meat, and even baked goods. It is now possible to get almost all one's food, locally produced, by specialized growers who support one another through integrated networks. This is an important model because, as oil becomes scarce and large-scale industrial farms struggle, smaller, integrated, localized farms will be critical.

As demand outpaces the ability to extract oil and natural gas, the prices of these commodities will steeply rise. We are already experiencing the beginnings of this trend. At what price will these resources become so expensive that the feedback from the increased costs will start to cripple our current economic system? Today 70 percent of the world's oil goes to transportation, and at present we don't have alternative energy resources to replace oil needed to fuel ships, jets, trucks, diesel trains, and cars. What will be the impact of decreased transportation capacity on the economic system?

In Keene, New Hampshire, where I teach and in Brattleboro, Vermont, close to where I live, citizens are being proactive and forming committees to analyze the implications of declining oil. They are examining potential problems due to decreased supplies and sustainable alternatives. These committees include sustainable designers, energy experts, transportation specialists, and community organizers—all working together to visualize integrated, efficient systems to support local communities as they transition away from cheap oil.

We also have a number of "sustainable communities" in our region. These are groups of people who intentionally live together in small communities dedicated to leading sustainable lifestyles. They grow much of their food, have energy efficient homes, share communal cars, and work to limit their collective environmental impact. They model the importance of connection to community and place.

Along with the increase in CSAs we have also witnessed the increase in small, locally owned, specialized commercial enterprises. I am confident that CSAs, sustainable communities, and local enterprises will continue to increase in our region. As they do, they will start creating the infrastructure for a sustainable economy. With rising energy costs, I also believe market dynamics eventually will allow these energy efficient farms and local businesses to sell their goods at lower prices than the huge oil-dependent corporations. We may witness a day when big box stores fail as it becomes increasingly difficult to transport goods globally. I can even imagine the removal of those stores, built on prime agricultural land, to make way for an increasing array of local farms. These sorts of trends provide hope for the future, as do the important guidance of spiritual teachers like Black Elk.

We need to follow their lessons. If we do, and work proactively toward sustainability in our personal and family lives as well as for change at the community, state, national, and even global levels, collectively we will become a critical part of the feedback that will change the system. And this change is essential for the well-being of our children, our grandchildren, numerous generations to come, as well as all life on this planet—which Black Elk saw as one united family.

"Is not the sky a father and the earth a mother, and are not all living things with feet or wings or roots their children? And this hide upon the mouthpiece here, which should be bison hide, is for the earth, from whence we came and at whose breast we suck as babies all our lives, along with all the animals and birds and trees and grasses. And because it means all this, and more than any man can understand, the pipe is holy."[1]

To work for strong connections within our community, to be grounded in our place, to support ourselves through reflective practice, and to create sustainable networks—I'm convinced Black Elk would see such endeavors as holy work.

Note

1. John Neihardt, 1932. *Black Elk Speaks* (Lincoln: University of Nebraska Press), 3.

Glossary of Scientific Terms

Anti-entropic: Also known as negentropic. Any open system that takes in more energy than it releases and stores that energy in growth and increasing complexity.

Beltian bodies: Small globules on the leaf tips of the Bull's-horn acacia that are harvested by acacia ants for protein.

Bifurcation: A rapid, large-scale change in a complex system's behavior due to iterative, positive feedback within the system.

Biodiversity: Biological diversity at any scale, but if the scale is not defined it is inferred to be at the species level, which is the number of species within an ecosystem. In this regard biodiversity is the same thing as species richness.

Biomass: The total amount of both living and nonliving organic matter in an organism or ecosystem.

Biosphere: The largest scale of biological organization, which includes all of the ecosystems that cloak the planet.

Butterfly effect: The idea that a small-scale change in a complex system through feedback can eventually result in large-scale consequences.

Carrying capacity: The maximal population size that an ecosystem can support without being degraded.

Cellular respiration: The breaking down of carbohydrate molecules within a cell to access the stored energy within the molecules.

Chaos theory: The original name given to the study of nonlinear systems. It is now called complex systems science.

Closed system: A system into which energy and materials cannot enter.

Coevolution: The process by which species adapt to each other so that they can more successfully coexist.

Competitive exclusion: The process by which a very competitive species excludes other species from an ecosystem.

Complex system: A nonlinear system that feeds back on itself as the parts interact in different ways at different times.

Complex systems science: The study of nonlinear systems. It used to be called chaos theory.

Convergence: The process in which minor perturbations within a system tend to cancel each other out.

Dynamic equilibrium: The state in an open system where the amount of energy entering the the system equals the amount leaving the system.

Ecosystem: An open system that includes populations of species, nonliving biomass, and the physical environment all interacting together.

Emergent properties: Aspects of a complex system's behavior that result from interactions within the system that couldn't have been predicted by an examination of the system's parts. Due to emergent properties, in a complex system the whole is greater than the sum of its parts.

Entelechy: Self-completing, Aristotle's view of the natural world. Matter and form are linked in a continuous process of change.

Entropy: A process in which a system becomes disorganized and simplified because it loses more energy from its transformations.

First law of thermodynamics: Law of conservation of energy that states that energy can neither be created nor destroyed.

Fuzzy boundaries: Spatial boundaries that allow for the movement of energy, materials, and information between nested complex systems.

Habitat: The ecosystem or ecosystems in which an organism lives.

Intermediate levels of disturbance: A disturbance to an ecosystem that is moderate and not large-scale.

Keystone predator: A predator that keeps competitive exclusion in check and therefore fosters high levels of species richness within an ecosystem.

Kinetic energy: The energy of motion.

Law of conservation of energy: The first law of thermodynamics, which states that energy can neither be created nor destroyed.

Law of entropy: The second law of thermodynamics, which states that when energy is transformed from one state to another, some of the energy is lost from the system where the transformation occurs, resulting in entropy.

Limits to growth: The law that states that all systems have to reach dynamic equilibrium in order to sustain themselves.

Linear system: A system in which all the parts work in a lockstep pattern, like a machine.

Microhabitat: A subportion of a habitat where an organism exists.

Mutualism: A mutually beneficial interaction between two individuals from different species that is essential for the survival of one or usually both individuals.

Mycorrhizae: Fungi that get their carbohydrate energy from the roots of plants while allowing the plants to dramatically increase their uptake of nutrients and water.

Negative feedback: Feedback within a complex system that maintains the system's status quo.

Nestedness: Complex systems at different spatial scales that are nested one within another.

Niche: The totality of all an organism's interactions with other organisms and the physical environment; its total ecological role.

Nonlinear system: A complex system that feeds back on itself as the parts interact in different ways at different times.

Old-growth forest: A forest that has reached dynamic equilibrium.

Open system: A system in which energy and materials flow in and out.

Photosynthesis: The cellular process by which light energy is used to combine carbon dioxide and water, forming carbohydrate energy storage molecules.

Positive feedback: Directional, iterative feedback that may bring about a bifurcation event.

Proximate knowledge of initial conditions: The idea that absolutely accurate measurements are not necessary for making accurate predictions.

Punctuated equilibria: The theory that the development of new species occurs quickly, not gradually, and is followed by long periods of stasis where little further change occurs.

Reductionism: The scientific approach that complex phenomena can be understood by examining their parts.

Saprophytes: Decay-producing organisms that get their energy from nonliving organic material.

Second law of thermodynamics: The law of entropy, which states that when energy is transformed from one state to another, some of the energy is lost from the system where the transformation occurs, resulting in entropy.

Self-organization: Increasing complexity in a complex system as it takes in more energy than it releases.

Species richness: The number of species in an ecosystem.

Static equilibrium: The final entropic state of a closed system that has dissipated its energy.

Succession: Changes in populations of species in an ecosystem through time, initiated by some form of disturbance.

Uniformitarianism: Hutton's theory that all geological features could be explained by slow, accumulative change.

Index